Advances in Experimental Medicine and Biology

Neuroscience and Respiration

Volume 921

More information about this series at http://www.springer.com/series/13457

Mieczyslaw Pokorski
Editor

Allergy and Respiration

Springer

Editor
Mieczyslaw Pokorski
Public Higher Medical Professional School in Opole
Institute of Nursing
Opole, Poland

ISSN 0065-2598 ISSN 2214-8019 (electronic)
Advances in Experimental Medicine and Biology
ISBN 978-3-319-42003-5 ISBN 978-3-319-42004-2 (eBook)
DOI 10.1007/978-3-319-42004-2

Library of Congress Control Number: 2016948845

Printed on acid-free paper

This Springer imprint is published by Springer Nature
The registered company is Springer International Publishing AG Switzerland

Preface

The book series *Neuroscience and Respiration* presents contributions by expert researchers and clinicians in the field of pulmonary disorders. The chapters provide timely overviews of contentious issues or recent advances in the diagnosis, classification, and treatment of the entire range of pulmonary disorders, both acute and chronic. The texts are thought as a merger of basic and clinical research dealing with respiratory medicine, neural and chemical regulation of respiration, and the interactive relationship between respiration and other neurobiological systems such as cardiovascular function or the mind-to-body connection. The authors focus on the leading-edge therapeutic concepts, methodologies, and innovative treatments. Pharmacotherapy is always in the focus of respiratory research. The action and pharmacology of existing drugs and the development and evaluation of new agents are the heady area of research. Practical, data-driven options to manage patients will be considered. New research is presented regarding older drugs, performed from a modern perspective or from a different pharmacotherapeutic angle. The introduction of new drugs and treatment approaches in both adults and children also is discussed.

Lung ventilation is ultimately driven by the brain. However, neuropsychological aspects of respiratory disorders are still mostly a matter of conjecture. After decades of misunderstanding and neglect, emotions have been rediscovered as a powerful modifier or even the probable cause of various somatic disorders. Today, the link between stress and respiratory health is undeniable. Scientists accept a powerful psychological connection that can directly affect our quality of life and health span. Psychological approaches, by decreasing stress, can play a major role in the development and therapy of respiratory diseases.

Neuromolecular aspects relating to gene polymorphism and epigenesis, involving both heritable changes in the nucleotide sequence and functionally relevant changes to the genome that do not involve a change in the nucleotide sequence, leading to respiratory disorders will also be tackled. Clinical advances stemming from molecular and biochemical research are but possible if the research findings are translated into diagnostic tools, therapeutic procedures, and education, effectively reaching physicians and patients. All that cannot be achieved without a multidisciplinary, collaborative, bench-to-bedside approach involving both researchers and clinicians.

The societal and economic burden of respiratory ailments has been on the rise worldwide leading to disabilities and shortening of life span. COPD alone causes more than three million deaths globally each year. Concerted efforts are required to improve this situation, and part of those efforts are gaining insights into the underlying mechanisms of disease and staying abreast with the latest developments in diagnosis and treatment regimens. It is hoped that the books published in this series will assume a leading role in the field of respiratory medicine and research and will become a source of reference and inspiration for future research ideas.

I would like to express my deep gratitude to Mr. Martijn Roelandse and Ms. Tanja Koppejan from Springer's Life Sciences Department for their genuine interest in making this scientific endeavor come through and in the expert management of the production of this novel book series.

Opole, Poland Mieczyslaw Pokorski

Contents

Advs Exp. Medicine, Biology - Neuroscience and Respiration (2016) 24: 1–10
DOI 10.1007/5584_2015_204

Published online: 29 January 2016

Spirometry or Body Plethysmography for the Assessment of Bronchial Hyperresponsiveness?

R. Merget, F. Nensa, E. Heinze, D. Taeger, and T. Bruening

Abstract

Methacholine testing is one of the standard tools for the diagnosis of mild asthma, but there is little information about optimal outcome measures. In this study a total of 395 college students were tested by the ATS dosimeter protocol for methacholine testing, with minor modification. Body plethysmography and spirometry were measured after each inhalation step. The end-of-test-criteria were (i) decrease in forced expiratory volume in 1 s (FEV_1) of $\geq$ 20 % and (ii) doubling of specific airway resistance and its increase to $\geq$ 2.0 kPa·s. The results were expressed by receiver operating characteristic (ROC) plots using questionnaire answers as a reference. The areas under the ROC curves were iteratively calculated for a wide range of thresholds for both measures. We found that ROC plots showed maximal sensitivities of about 0.5–0.6 for FEV_1 and about 0.7 for specific airway conductance (sGt), with similar specificities of about 0.7–0.8 taking questions with the known high specificity as references. Accordingly, larger maximal areas under the ROC curve were observed for body plethysmography, but the differences were small. A decrease in FEV_1 of about 15 % and a decrease of sGt of about 60 % showed the largest areas under the ROC curves. In conclusion, body plethysmography yielded better sensitivity than spirometry, with similar specificity. However, replacing the common spirometric criterium for a positive test (20 % decrease in FEV_1 from baseline) by the optimal body plethysmographic criterium would result in an increase of false positive tests from about 4 to 8 % in healthy young adults.

R. Merget (✉), F. Nensa, E. Heinze, D. Taeger, and T. Bruening
Institute for Prevention and Occupational Medicine of the German Social Accident Insurance, Institute of the Ruhr University, 1 Bürkle-de-la-Camp-Platz, 44789 Bochum (IPA), Germany
e-mail: merget@ipa-dguv.de

Keywords
Body plethysmography • Bronchial hyperresponsiveness • Dosimeter • Methacholine • Spirometry

1 Introduction

The assessment of bronchial hyperresponsiveness to methacholine or histamine is one of the standard tools for the diagnosis of mild asthma. Change in forced expiratory volume in one second (FEV_1) has been chosen as the primary outcome measure for methacholine testing by the American Thoracic Society (ATS). According to the ATS guidelines, body plethysmography should be used primarily in patients who cannot perform acceptable spirometry maneuvers. The guidelines suggest a decrease of specific airway conductance of 45 % (ATS 2000) as a threshold. The ATS recommendation is based on a better reproducibility compared with other indices (Dehaut et al. 1983) and a better discrimination of asthmatic subjects from normals (Cockcroft and Berscheid 1983; Michoud et al. 1982).

There is good evidence that body plethysmography has a higher sensitivity than spirometry (Khalid et al. 2009; Goldstein et al. 1994; Cockcroft and Berscheid 1983; Dehaut et al. 1983). However, all those studies have been performed as case-control studies with a limited number of subjects and a limited number of thresholds for both measures. With a case-control design it is highly probable that the test with higher specificity will perform better than a more sensitive test that will be positive also in a few healthy controls. Accordingly, spirometry may be not sensitive enough, which would result in a number of patients who have asthma, but a negative test.

A further important point which may be criticized in those studies is the threshold of the airway reaction that defines a positive test. Almost all studies used a very limited number of thresholds or terminated the test after one single threshold was observed, and also data were retrospectively analyzed. In a recent study, we tried to overcome these shortcomings by analyzing a wide range of thresholds (Nensa et al. 2013). As a result, body plethysmography demonstrated higher sensitivity with comparable specificity. We also used a case-control design, but the subjects were recruited from the cases of possible occupational asthma, which may have biased the results. In the present study, a large cohort of college students were examined with both measures, self-reported symptoms were used as a reference, and a wide range of thresholds was compared.

2 Methods

The study was approved by the Ethics Committee of the Ruhr-University in Bochum, Germany (permit no. 1555) and all subjects gave their written informed consent.

2.1 Subjects

Within three years, a total of 829 young subjects (median 25, range 20–40 years) were recruited from medical university students and asked to participate in this study. Of these, 749 (90.3 %) agreed to answer a questionnaire regarding respiratory symptoms and to perform both body plethysmography and spirometry. Contraindications for methacholine (MCH) testing were present in 145 subjects (acute bronchitis within the previous 6 weeks: n = 122, current asthma medication: n = 8, pregnancy: n = 3, poor breathing technique: n = 8, airway obstruction after spirometry: n = 4). Of the 604 remaining subjects, 173 did not agree to methacholine testing and 36 subjects completed the test, but were excluded due to insufficient cooperation (see below). Thus the study population comprised 395 students (47.6 % of those initially recruited) who showed acceptable MCH tests.

2.2 Study Protocol

After agreement for participation, subjects answered the questionnaire and performed lung function tests. If no contraindication was present, MCH testing followed. Between 1 and 4 subjects per day were invited to perform the tests. Measurements were done between 1 and 4 p.m. in an air conditioned room (temperature 24 °C, relative humidity 50 %). Two trained technicians performed all tests, and the first author was present during all challenges, with very few exceptions.

2.3 Questionnaire

The questionnaire recommended by the ATS (ATS 2000) was used. It asks for smoking status, determinants of disease, and for contraindications such as respiratory infection in the last 6 weeks (translated into 'acute bronchitis (i.e., cough and phlegm) in the last 6 weeks), pregnancy, high blood pressure, heart attack or stroke within the last 3 months, aortic aneurysm, and medication). In addition, subjects were asked about wheezing during the last year and wheezing ever after age 18. A summary of the questionnaire items that were used as references are shown in Table 1.

2.4 Body Plethysmography and Spirometry

The same equipment (Masterscreen; CareFusion, Höchberg, Germany) was used throughout the study. Calibration was performed daily. The breathing maneuvers performed at baseline and during methacholine testing were standardized as follows: total specific airway resistance (sR_t) was determined from five reproducible pressure-flow curves during normal quiet breathing, whereby the maximal in- and end-expiratory pressures were used. Specific airway conductance (sG_t) was calculated from sR_t. After having recorded the pressure-flow loops, the airway shutter was closed at end-expiration and functional residual capacity (FRC) determined according to current recommendations (Criée et al. 2011). Immediately afterwards, subjects exhaled to residual volume (RV), inhaled quickly to total lung capacity (TLC) and performed a forced expiration. Spirometry was done as recommended by the ATS (ATS 1995) and the ECCS reference values were used for the normalization of data (Quanjer et al. 1993).

2.5 Methacholine Testing

Methacholine (Synopharm; Barsbüttel, Germany) was prepared as a 32 mg/mL stock solution using 0.9 % saline. This solution was stored no longer than for 4 weeks at 4 °C; dilutions were done weekly. The APSpro

Table 1 Questionnaire items used as references

Item No.	Item description of ATS	Item description of this study
1	Has a physician told you that you have asthma?	As ATS
2	Have you ever been hospitalized for asthma?	As ATS
3	Did you have respiratory disease as a child?	Did you have recurrent or chronic respiratory disease as a child?
4	Have you ever experienced asthma symptoms such as wheezing, chest tightness, or shortness of breath within the last two weeks?	As ATS
5	–	Did you experience wheezing within the last year?
6	–	Did you experience wheezing at least once after age 18?

dosimeter (CareFusion, Höchberg, Germany) and DeVilbiss 646 nebulizer (DeVilbiss; Malsch, Germany; the same throughout the study) were utilized for nebulization. The nebulizer was filled with 2 mL of the solution. Its straw had been fixed by gluing (Jörres et al. 1992) to exclude alterations of its position as the source of the previously reported variability (Hollie et al 1991).

The ATS protocol comprises 5 concentrations (0.0625, 0.25, 1.0, 4.0, 16.0 mg/mL) administered in 5 consecutive steps, without initial inhalation of the diluent. Each of these concentrations was given in 5 consecutive slow inspirations from FRC to near TLC, while the nebulizer was actuated over 0.6 s. Inspiratory airflow was kept close to 1 L/s by observing a visual scale. The time interval between consecutive inhalation steps was about 5 (range 4–6) min. The ATS protocol was implemented with two minor modifications: (1) no breathhold was performed after inhalation, (2) nebulizations were initiated 0.5 s after the beginning of each inhalation to achieve a significant airflow upon nebulization.

Body plethysmography was performed 60 s after the last of the five consecutive inhalations of each step. This implied that spirometry was done about 90 s after inhalation. If spirometry had to be repeated, this was done without previous body plethysmography. For baseline measurements and after the last step of methacholine testing, three acceptable spirometric maneuvers were required, while the inhalation steps in-between were followed by only one acceptable maneuver to reduce the potential impact of repeated forced expirations. Tests were terminated if either the last dose had been reached or a fall of FEV_1 of ≥ 20 % (FEV_1-criterion) and a fall of sG_t of ≥ 60 % to ≤ 0.5 $(kPa \cdot s)^{-1}$ (sG_{t60+}) (sGt-criterion) was documented. We chose sG_t instead of sR_t for data evaluation to ensure the same direction of changes in the functional indices.

2.6 Quality Control

Output determinations of the combination of pressure generator and nebulizer as used in the challenges were performed at a pressure of 1.3×10^5 Pa and a flow of 9.6 L/min. The delivered output ranged within the limits of 900 ± 90 mg/min given by the ATS. Outputs were assessed weekly by weighing. In doing this, a control subject breathed through the mouthpiece, whereby both inhalation and exhalation took place through the nebulizer. There were no trends toward increasing or decreasing output over time. For FEV_1 and forced vital capacity the criteria of reproducibility as described by the ATS were used. To further improve the quality of spirometric data, subjects who showed a ≥ 5 % increase of FEV_1, compared with the previous measurement, after any MCH dose were excluded from the analysis ($n = 36$), as this was considered to reflect insufficient cooperation.

2.7 Data Analysis

Taking the questionnaire data as reference, receiver operating characteristic (ROC) plots were established (Zweig and Campbell 1993). This was done separately for tests positive with regard to FEV_1 and tests positive with regard to sG_t. We calculated ROC plots with the positivity criterium for FEV_1 (20 % decrease) and for sG_t with a decrease to ≤ 0.5 $(kPa \cdot s)^{-1}$ and thresholds with decreases ≥ 60 % (sG_{t60+}), ≥ 50 % (sG_{t50+}) or ≥ 40 % (sG_{t40+}) of baseline. In order to be able to analyze the whole data set, we preferred dose-response slopes over provocative concentration (PC) or provocative dose (PD) values, as many tests were negative with regard to the corresponding criteria. The slopes of the dose-response curves for both end-of-test-criteria (FEV_1 and sG_t) were calculated as described earlier (O'Connor et al. 1987). Analyses were done by linear regression of log-transformed data. Furthermore, questionnaire data and baseline lung function were compared between subjects with ($n = 395$) and without ($n = 354$) accepted methacholine tests. For the comparison of groups, Mann-Whitney U test or Fisher's exact test were used. The correlations between the dose-response slopes of sG_t and those of FEV_1 were calculated with Spearman's rank correlation test. Statistical significance was assumed if $p < 0.05$. All analyses were performed with SAS 9.2 (Cary, NC).

3 Results

The mean normal FEV_1 values characterize an overall healthy cohort. The subjects with acceptable tests did not differ from those who were not included in the final analysis, with the exception that men were more likely to perform the tests and to fulfil the quality criteria (Table 2).

Wheezing, chest tightness, or shortness of breath within the last two weeks (Question 4) was answered positively in 22 (5.6 %) cases and yielded the steepest slopes of the dose-response curves for both FEV_1 and sG_t, which allowed a good separation of those who answered all questions negative, with little overlap (Fig. 1). The slopes for FEV_1 and sG_t showed overall high correlation (Fig. 2). ROC plots showed maximal sensitivities of about 0.5–0.6 for FEV_1 and about 0.7 for sGt (Questions 1, 2, and 4) with roughly similar specificities of about 0.7–0.8 (Fig. 3).

Larger maximal areas under the ROC curve were observed for body plethysmography, but the differences were small, with the highest value obtained with the combination of Questions 1 and 5 (Fig. 4). In this cohort of young college students with an overall low pre-test probability for a positive test, the result was dominated by the second body plethysmographic criterium of a decrease of sGt to ≤ 0.5 $(kPa{\cdot}s)^{-1}$. Decreases of FEV_1 of about 15 % and of sGt of about 60 % were optimal (Fig. 4).

For the whole cohort, the optimal body plethysmographic criterium was compared with the common spirometric criterium. In completely healthy subjects, the number of positive tests after step 4, which would be considered positive by the ATS was 3.6 % with the FEV_1 criterium and 8 % with the sGt criterium (Table 3).

Table 2 Comparison of baseline data of participants who answered the questionnaire and performed lung function tests ($n = 749$) according to availability of methacholine testing of sufficient quality (see Methods)

	Methacholine test		p-value
	Yes (n = 395)	No (n = 354)	
Age (years; median, min-max)	25 (20–39)	25 (22–40)	0.10
Female gender (n; %)	196 (49.6)	206 (58.0)	0.02
Questionnaire			
Smoking: current/ex (n; %)	84 (22.2)	98 (28.2)	0.06
Question 1* (n; %)	46 (11.7)	30 (8.5)	0.18
Question 2* (n; %)	5 (1.3)	3 (0.9)	0.73
Question 3* (n; %)	90 (22.8)	77 (21.8)	0.79
Question 4* (n; %)	22 (5.6)	23 (6.5)	0.65
Question 5* (n; %)	58 (14.7)	44 (12.4)	0.39
Question 6* (n; %)	87 (22.0)	76 (21.5)	0.93
Question 1 or 2 or 4* (n; %)	56 (14.2)	42 (11.9)	0.39
Question 1 and 5* (n; %)	30 (7.6)	23 (6.5)	0.57
Answers to all questions negative (n; %)	251 (63.5)	228 (64.4)	1.0
Baseline lung function			
FEV_1 (% pred; median, min-max)	105.9 (73.9–140.7)	104.9 (62.9–139.6)	0.31
sGt $((kPa{\cdot}s)^{-1}$; median, min-max)	1.45 (0.57–4.24)	1.46 (0.52–2.94)	0.78

*Question numbers refer to Table 1; p-values: Fisher's exact test for categorical data and Wilcoxon's rank-sum for baseline lung function and age

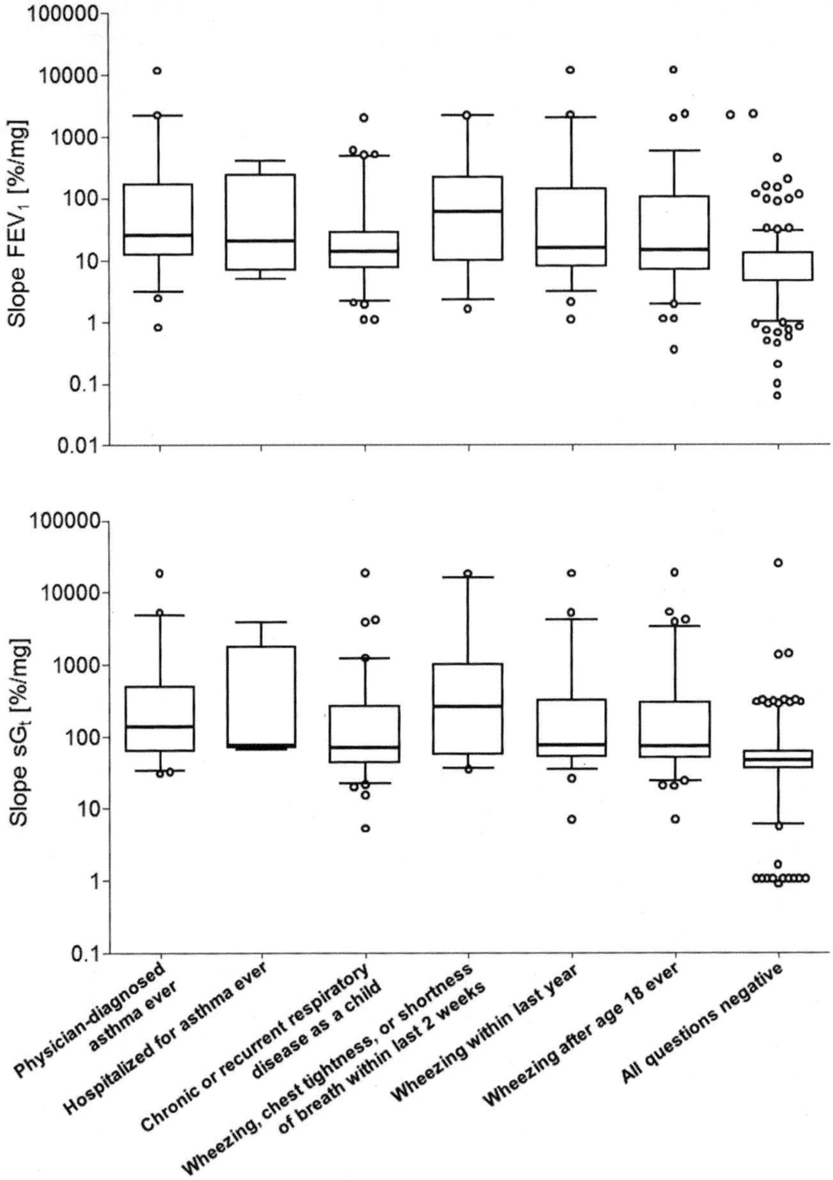

Fig. 1 Dose-response slopes of methacholine (MCH) tests depicted on a log scale as assessed with FEV_1 (A) and specific airway conductance (sG_t) (B) for different items of the ATS questionnaire (see Table 1). *Box-plots* indicate the geometric mean, 5 %, 25 %, 75 %, and 95 % quantiles. *Dots* represent the individuals below the 5 % and above the 95 % quantiles. The number of subjects with positive answers is given in Table 2

4 Discussion

Body plethysmography is widely used in Germany for the assessment of bronchial hyperresponsiveness (Criée et al. 2011). It has the advantage that no deep inspiration is necessary, which has been demonstrated to affect bronchial hyperresponsiveness (Cockcroft and Davis 2006). In addition, spirometry asthma can be avoided, and the result is less dependent of the subjects' cooperation. Finally, as no forced maneuvers are required, cough that may hinder the test interpretation is minimized.

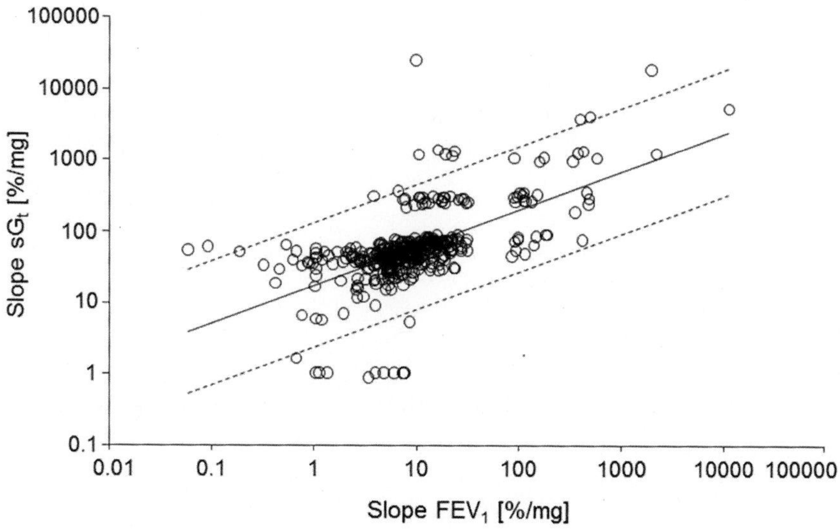

Fig. 2 Dose-response slopes of specific airway conductance (sG_t) and FEV_1. Both data sets are given on a log scale. The orthogonal linear regression and 95 % CI lines are also shown. Overall, Spearman's correlation coefficient was 0.64 (95 % CI 0.58–0.70)

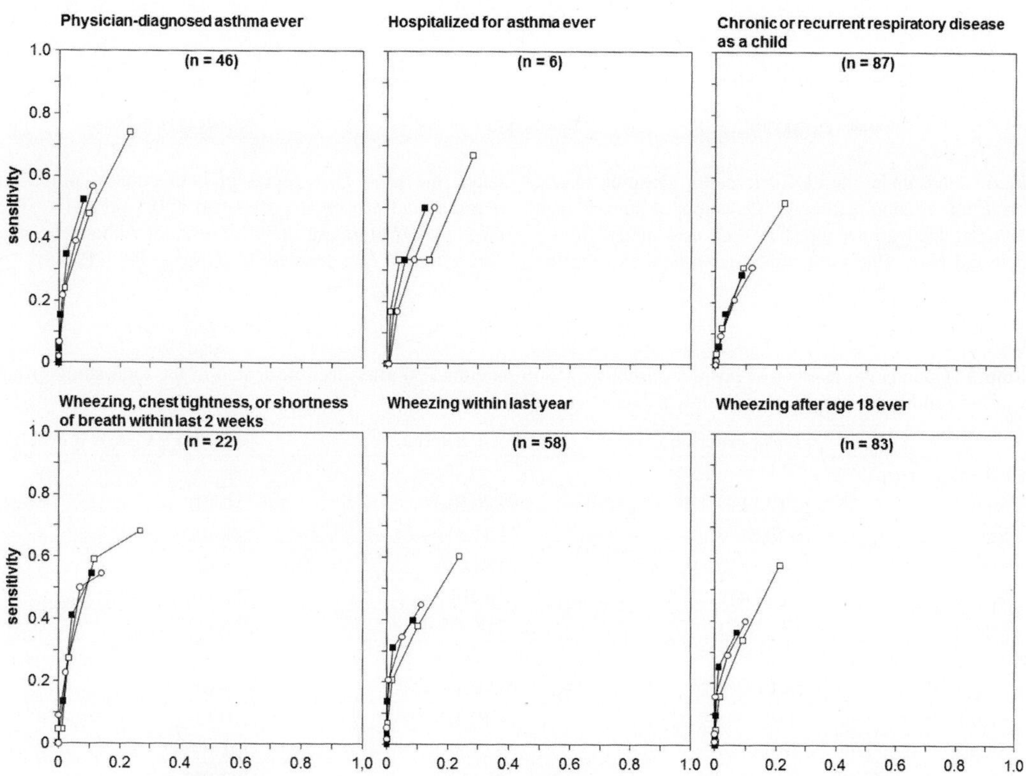

Fig. 3 ROC plots with the answers to the ATS questionnaire as reference. *Open dots* indicate tests which were positive with the FEV_1 criterion, filled squares for the sG_{t60+} criterion, and *open squares* with the sG_{t40+} criterion (see Methods). The sG_{t50+} criterion is identical with the sG_{t60+} criterion and not shown in this figure

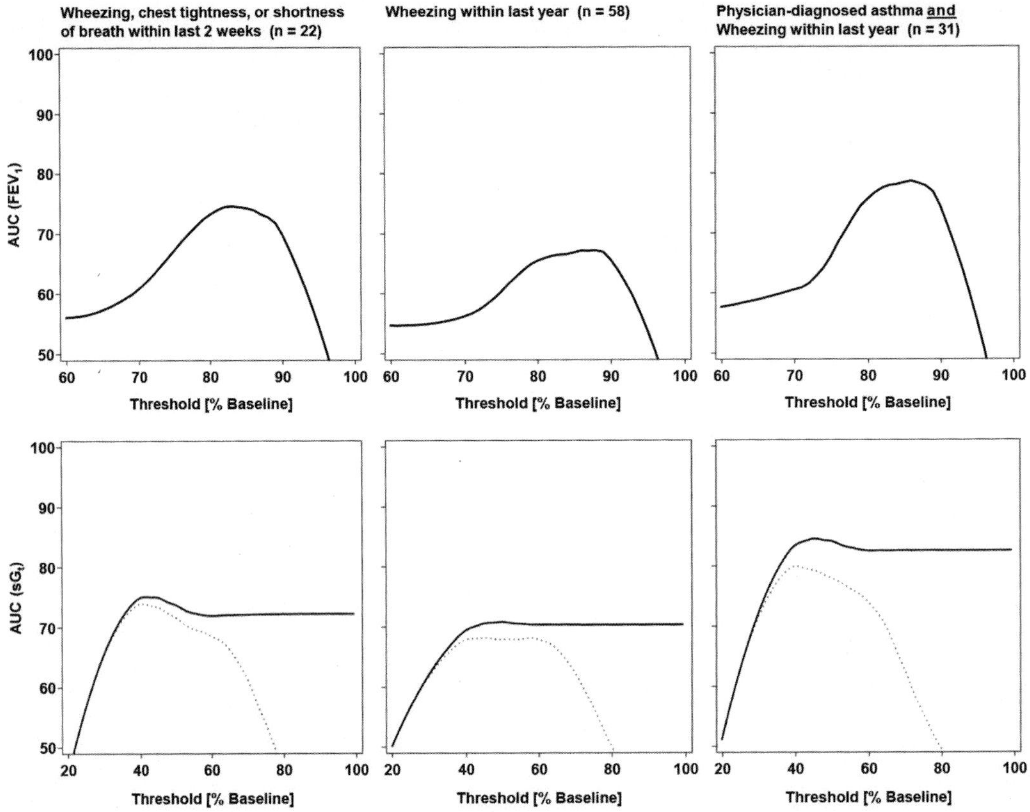

Fig. 4 Areas under the ROC curves for Questions 4 and 5 and for a combination of Questions 1 and 5 (see Methods) for various thresholds of spirometry (*upper part*) und body plethysmography (*lower part*). For the latter, the *solid lines* represent both criteria, fall from baseline and a decrease of sGt to $\leq 0.5\ (\text{kPa·s})^{-1}$. The *dashed lines* represent the sGt criterium without the second criterium of a decrease of sGt to $\leq 0.5\ (\text{kPa·s})^{-1}$

Table 3 Cumulative number of positive responses (% in parentheses) after each dose step in the total study group (n = 395) and in the subgroup of healthy subjects[c] (n = 251)

	FEV_1 criterion[a]	sGt criterion[b]	Both FEV_1 & sGt criteria
Total study group			
Step 1	1 (0.3)	2 (0.5)	0 (0)
Step 2	3 (0.8)	5 (1.3)	2 (0.5)
Step 3	14 (3.5)	18 (4.6)	9 (2.3)
Step 4	37 (9.4)	56 (14.2)	25 (6.3)
Step 5	64 (16.2)	116 (29.4)	53 (13.4)
Healthy subjects[c]			
Step 1	0 (0)	1 (0.4)	0 (0)
Step 2	0 (0)	1 (0.4)	0 (0)
Step 3	1 (0.4)	3 (1.2)	1 (0.4)
Step 4	8 (3.2)	15 (6.0)	2 (0.8)
Step 5	23 (9.2)	43 (17.1)	14 (5.6)

[a]Fall of $FEV_1 \geq 20\ \%$ from baseline
[b]Fall of specific airway conductance (sG_t) $\geq 60\ \%$ from baseline to $\leq 0.5\ (\text{kPa·s})^{-1}$
[c]Subjects who answered all questions concerning asthma symptoms negative

The largest disadvantages are higher costs, no availability in epidemiologic field studies, or claustrophobia. The latter point has not been addressed in the present study. Instead, we intended to answer the question how sensitivities and specificities of both methods vary using a wide range of thresholds as positivity criteria. That has never been done in a cohort study including subjects with symptoms, but no clear diagnosis of asthma.

The main finding of this study is a higher sensitivity of body plethysmography without a relevant loss of specificity. This implies that, depending on the reference, the commonly used spirometric positivity criterion yields about 20 % false negative tests, but the optimal body plethysmographic criterion increases the rather low number of false positive tests from about 4–8 %. The number of 8 % of false positive tests with body plethysmography may be considered acceptable, as it nears the 5th percentile, which is a common cut-off in lung function reference equations. This has to be weighed against the relevant decrease of false negative spirometric tests.

This study has restricted the analysis to absolutely high quality spirometry, which will possibly not be reached in practice or in subjects with a higher pre-test probability for a positive test. However, similar results were obtained recently with a case-control design (Nensa et al. 2013). Thus, the results should be valid also in subjects with higher asthma severity. If the quality of spirometry is not as high as in this study, the result would be a higher number of false positive tests. As this is not a major problem of spirometry as an outcome measure, our results should be transferable to daily practice.

The results of the present study indicate that there is a substantial underdiagnosis of bronchial hyperresponsiveness in epidemiological studies. This cannot be overcome by lowering the spirometric 20 % threshold to 15 %, as both thresholds yield similar results.

Another interesting finding of this study is a low sensitivity of the tests with all questions as reference, i.e., a considerable number of false 'negative' tests. This has to be considered in epidemiologic studies relying on questionnaire data which possibly overestimate the existence of bronchial hyperresponsiveness and asthma.

In summary, the more costly body plethysmography has several advantages for the assessment of bronchial hyperresponsiveness. Whether better reproducibility of spirometry outweighs these advantages needs further study and a critical choice depending on the aim of testing.

Conflicts of Interest The authors declare no conflicts of interest in relation to this study.

References

ATS- American Thoracic Society (1995) Standardization of spirometry (1994 update). Am J Respir Crit Care Med 152:1107–1136

ATS- American Thoracic Society (2000) Guidelines for methacholine and exercise challenge testing-1999. Am J Respir Crit Care Med 161:309–329

Cockcroft DW, Berscheid BA (1983) Measurement of responsiveness to inhaled histamine: comparison of FEV_1 and sGaw. Ann Allergy 51:374–377

Cockcroft DW, Davis BE (2006) The bronchoprotective effect of inhaling methacholine by using total lung capacity inspirations has a marked influence on the interpretation of the test result. J Allergy Clin Immunol 117:1244–1248

Criée CP, Sorichter S, Smith HJ, Kardos P, Merget R, Heise D, Berdel D, Köhler D, Magnussen H, Marek W, Mitfessel H, Rasche K, Rolke M, Worth H, Jörres RA (2011) Body plethysmography - its principles and clinical use. Respir Med 105:959–973

Dehaut P, Rachiele A, Martin RR, Malo JL (1983) Histamine dose–response curves in asthma: reproducibility and sensitivity of different indices to assess response. Thorax 38:516–522

Goldstein MF, Pacana SM, Dvorin DJ, Dunsky EH (1994) Retrospective analyses of methacholine inhalation challenges. Chest 105:1082–1088

Hollie MC, Malone RA, Skufca RM, Nelson HS (1991) Extreme variability in aerosol output of the DeVilbiss 646 jet nebulizer. Chest 100:1339–1344

Jörres R, Nowak D, Rabe K, Magnussen H (1992) Variability in aerosol output of the DeVilbiss 646 jet nebulizer. Chest 102:1636

Khalid I, Morris ZQ, Digiovine B (2009) Specific conductance criteria for a positive methacholine challenge test: are the American Thoracic Society guidelines rather generous? Respir Care 54:1168–1174

Michoud MC, Ghezzo H, Amyot R (1982) A comparison of pulmonary function tests used for bronchial challenges. Bull Eur Physiopathol Respir 18:609–621

Nensa F, Kotschy-Lang N, Smith HJ, Marek W, Merget R (2013) Assessment of airway hyperresponsiveness: Comparison of spirometry and body plethysmography. Adv Exp Med Biol 755:1–9

O'Connor G, Sparrow D, Taylor D, Segal M, Weiss S (1987) Analysis of dose response curves to methacholine. An approach suitable for population studies. Am Rev Respir Dis 136:1412–1417

Quanjer PH, Tammeling GJ, Cotes JE, Pedersen OF, Peslin R, Yernault JC (1993) Lung volumes and forced ventilatory flows. Report working party standardization of lung function tests, European community for steel and coal. Eur Respir J 6(Suppl 16):5–40

Zweig MH, Campbell G (1993) Receiver operating characteristic (ROC) plots: a fundamental evaluation tool in clinical medicine. Clin Chem 39:561–577

Advs Exp. Medicine, Biology - Neuroscience and Respiration (2016) 24: 11–20
DOI 10.1007/5584_2016_242

Published online: 10 May 2016

Clinical Effects, Exhaled Breath Condensate pH and Exhaled Nitric Oxide in Humans After Ethyl Acrylate Exposure

F. Hoffmeyer, J. Bünger, C. Monsé, H. Berresheim, B. Jettkant, A. Beine, T. Brüning, and K. Sucker

Abstract

Ethyl acrylate is an irritant known to affect the upper airways and eyes. An increase of the eye blink frequency in humans was observed during exposure to 5 ppm. Studies on the lower airways are scant and our study objective was the evaluation of pH in exhaled breath condensate (EBC-pH) and nitric oxide in exhaled breath (FeNO) as markers of inflammation. Sixteen healthy volunteers were exposed for 4 h to ethyl acrylate at a concentration of 5 ppm and to sham (0.05 ppm) in an exposure laboratory. Clinical irritation symptoms, EBC-pH (at a pCO_2 of 5.33 kPa) and FeNO were assessed before and after exposure. Differences after ethyl acrylate exposure were adjusted for those after sham exposure. 5 ppm ethyl acrylate induced clinical signs of local irritation in the nose and eyes, but not in lower airways. Exposure produced a subtle, but statistically significant, decrease in breathing frequency (1 breath/min; $p = 0.017$) and a lower EBC-pH (by 0.045 units; $p = 0.037$). Concerning FeNO, we did not observe significant changes compared to sham exposure. We conclude that local effects induced by 5 ppm ethyl acrylate consist of sensory irritation of eyes and nose. In addition, acute ethyl acrylate exposure to 5 ppm resulted in a net decrease of EBC-pH. Whether that can be interpreted in terms of additional lower airway irritation or already inflammatory alterations set in needs further investigations.

Keywords

Acid-base balance • Breathing frequency • Ethyl acrylate • Exhaled nitric oxide • Sensory irritation

F. Hoffmeyer (✉), J. Bünger, C. Monsé, H. Berresheim, B. Jettkant, A. Beine, T. Brüning, and K. Sucker
Institute for Prevention and Occupational Medicine of the German Social Accident Insurance, Institute of the Ruhr-University Bochum (IPA), Bürkle-de-la-Camp-Platz1, 44789 Bochum, Germany
e-mail: hoffmeyer@ipa-dguv.de

1 Introduction

Ethyl acrylate is an ester of acrylic acid, used as a building block in polymer production in the paper, leather, and textile industries. It belongs to the organic chemicals which can be odorous (olfactory stimulation) and irritating (trigeminal, glossopharyngeal, and vagal stimulation). Sensory irritations are endpoints of effect and crucial for the regulation of many chemicals used in workplaces (Brüning et al. 2014; Arts et al. 2006). Ethyl acrylate is irritating to the skin and mucous membranes of the eyes, as well as to upper and lower airways. Concerning occupational exposure limits, the German MAK value ('Maximale Arbeitsplatz-Konzentration') of 5 ppm for ethyl acrylate (MAK 2012) was based on the irritation-induced histopathological changes in the nasal mucosa of rodents seen at higher concentrations (Miller et al. 1985). Olfactory receptors respond to ethyl acrylate at lower concentrations and with greater selectivity than the trigeminal nerve endings do. In previous work, the odor threshold of ethyl acrylate could be determined at 0.0066 ppb and the irritation threshold at 4.15 ppm (van Thriel et al. 2006). The irritation threshold, also referred to as the 'lateralization' threshold, is based on the ability to correctly localize trigeminal stimuli to the stimulated nostril, while olfactory stimuli cannot be lateralized (Kobal et al. 1989). In general, odor thresholds are lower than irritation thresholds (Cometto-Muñiz and Cain 1990). Metabolic studies in rats demonstrate that ethyl acrylate is rapidly absorbed, degraded mostly by hydrolysis to acrylic acid and ethanol and finally exhaled as CO_2. Complaints such as burning, dry, and itching eyes are among the most common symptoms. Changes in eye blink frequency as a measure of trigeminal stimulation were observed in a human challenge study with 5 ppm ethyl acrylate (Blaszkewicz et al. 2010) and recently, a lower MAK value (2 ppm) (MAK 2012) was recommended. Reaching the lower airways, ethyl acrylate at first contacts the airway lining fluid (ALF), where stimulation of peripheral nerves, and chemical events and reaction products promote biological responses in terms of pulmonary irritation and inflammation (Shusterman 2003). Short-term stimulation of the sensory irritation pathway is thought to be reversible. However, prolonged stimulation or higher concentrations might trigger a neurogenic inflammation cascade accelerating the general inflammatory defence mechanisms. These pathways may become indistinguishable when adverse health effects occur (Brüning et al. 2014; Arts et al. 2006).

Changes in cellular or biochemical markers of inflammatory responses could be assessed by more or less invasive techniques (Quirce et al. 2010). Inflammation in the context of several lung diseases is characterized by acidification of the airways (Kostikas et al. 2002). Exhaled breath condensate (EBC) reflects the ALF composition and the measurement of the pH in EBC is a valid method, especially when considering the influence of CO_2 in the analytic procedure (Hoffmeyer et al. 2015a; Kullmann et al. 2007). The fraction of exhaled nitric oxide (FeNO) reflects the activity of NO synthases induced during inflammation and its measurement is well standardized (ATS/ERS 2005). Apart from changes due to inflammatory processes, FeNO levels and pH could be interlinked and influenced by the composition of inhaled air. In this respect, smoking is a known confounder for detected levels of FeNO and EBC-pH (Koczulla et al. 2010; Kharitonov et al. 1995). In previous human challenge studies, we assessed acute effects of low dose sulfur dioxide and ozone exposure on the airways using non-invasive methods and demonstrated modulation of EBC-pH by exercise (Hoffmeyer et al. 2015b; Raulf-Heimsoth et al. 2010).

The objective of this study was to evaluate clinical effects, EBC-pH, and FeNO, as markers of airway inflammation, in healthy subjects after exposure to a concentration of ethyl acrylate (5 ppm) which is supposed not to affect lower airways.

2 Methods

The study was approved by a local Ethics Committee of the Ruhr University in Bochum Germany, and all study participants gave written informed consent. The protocol was created in accordance with the Declaration of Helsinki for Human Research.

2.1 Subjects

The 16 human volunteers were healthy non-smokers without airway sensitization. Sensitization to common inhalant allergens was evaluated using standard skin prick-tests. Smokers were excluded from the study. Smoking habits were assessed by face-to-face interviews and validated by quantification of the nicotine metabolite cotinine in urine.

Details on the methods used for functional characterization of the subjects have been previously described (Hoffmeyer et al. 2015b). Results of lung function variables and respective z-scores refer to predicted values derived from healthy non-smoking Caucasian subjects collected by the Global Lung Initiative (GLI) (Quanjer et al. 2012). There was no drop of at least 20 % in FEV_1 or a doubling of specific airway resistance within the four concentration steps of the methacholine challenge test in any subject. Study characteristics, including functional results, are summarized in Table 1.

Table 1 Subject characteristics

Gender, F/M, n	9/7
Age (year)	25 (23; 27)
BMI (kg/m^2)	21.5 (20.3; 23.8)
FEV_1 (%pred$_{GLI}$)	96.6 (90.3; 103.2)
z-score	−0.29 (−0.83; 0.26)
FVC (%pred$_{GLI}$)	100.8 (93.5; 110.1)
z-score	0.07 (−0.52; 0.82)
FEV_1/FVC (%pred$_{GLI}$)	96.4 (91.8; 100.0)
z-score	−0.50 (−1.00; 0.04)
$MEF_{25/75}$ (%pred$_{GLI}$)	86.7 (68.8; 100.5)
z-score	−0.62 (−1.43; 0.03)

F female, *M* male, *BMI* body mass index, *GLI* global lung initiative, FEV_1 forced expiratory volume in 1 s, *FVC* forced vital capacity, *MEF* $_{25/75}$ mean expiratory flow between 75 % and 25 % of vital capacity. Continuous variables are depicted with median and inter-quartile range (IQR)

2.2 Exposure

Details of the IPA exposure chamber have been reported elsewhere (Monsé et al. 2012). Subjects were exposed to either a constant ethyl acrylate concentration of 5 ppm or to 0.05 ppm (sham) for 4 h in a randomly blind cross-over design as previously reported (Hoffmeyer et al. 2015b). At 0.05 ppm, ethyl acrylate brought about an odor perception, but no mucosal irritation. Therefore, this concentration was chosen for sham exposure instead of filtered air to enable a blinded test design. Ethyl acrylate concentrations were monitored every 2 s *via* online mass spectroscopy using a chemical ionization mode (model airsense; MS4-Analysentechnik GmbH, Rockenberg, Germany).

2.3 Effect Assessment

Complaints, clinical symptoms, and biomarkers were assessed before ($_{pre}$) and immediately post exposure ($_{post}$).

Complaints and Clinical Symptoms Subjects were asked in an open-end manner, followed by a questionnaire on sensations or complaints of eyes, upper and lower airways. Irritations are perceived as pungency, stinging, and burning sensations. Sneezing and lacrimation characterize effects on the upper respiratory tract, cough and shortness of breath indicate effects on larynx or lower airways. Moreover, subjects were asked about fatigue, dizziness, headaches, and nausea. Results on annoyance and cognitive effects are not herein reported as they are judged a separate ramification of the study to be reported in a different study.

The subjects were examined for signs of conjunctivitis (watering eyes, tears, mucosal swelling, vascular injection), rhinitis (obstruction, mucosal swelling, vascular injection, rhinorrhea), throat irritation (mucosal swelling, vascular injection), and airway obstruction (auscultation). According to the overall intensity, complaints and clinical symptoms due to ethyl acrylate exposure were categorized into no (0), weak (1) moderate (2) or strong response (3).

Breathing Frequency Biosignals were recorded with a modular ambulatory polysomnography system SOMNOscreen™ Plus (PSG system), consisting of a PSG head box and a PC running the analysis and monitoring software Domino™ Ver. 2.6 (Somno Medics; Randsacker, Germany). Data from the device are directly wirelessly transferred and internally stored on a compact flash card (Smith et al. 2013).

Two piezo respiratory effort belts (abdominal and thoracic) were applied for the assessment of breathing frequency. After applying a low pass filter at 1 Hz on the breathing recordings, breathing frequency was calculated from the effort sum during the time intervals of 4.5 ($n = 11$) and 9 min ($n = 14$) to account for variability in breathing rate. For each subject, the median values of these 25 intervals were compared to evaluate exposure associated differences in breathing rate. Finally, the overall medians for each exposure condition were compared.

Biomarkers EBC was sampled during tidal breathing through a mouthpiece according to the general methodological recommendations with the temperature-controlled device Turbo DECCS (Medivac; Parma, Italy) (Horváth et al. 2005). Subjects were instructed to swallow excess of saliva after coming off the mouthpiece. The collection time was exactly 10 min at a maintained temperature of –5 °C. A blood gas analyzer (ABL800; Radiometer GmbH, Willich, Germany) was used for simultaneous determination of the pH and the partial pressure of carbon dioxide (pCO_2). Calibration solutions pH5, pH6, pH7, and pH8 were analyzed for the quality assessment. PCO_2 is the most important confounder of pH measurement in EBC samples. Therefore, we adjusted pH to a PCO_2 of 5.33 kPa ($pH_{5.33}$) as previously recommended (Hoffmeyer et al. 2015a; Kullmann et al. 2007).

FeNO was measured using a portable electrochemical analyzer (NIOX Mino; Aerocrine, Solna, Sweden) taking into account a guideline of the American Thoracic Society and European Respiratory Society (ATS/ERS 2005).

2.4 Statistical Elaboration

pH and FeNO were calculated as the percent change after exposure compared to start of exposure [(post-pre/pre)*100] and reported as ΔpH and ΔFeNO, respectively. Differences in biomarkers after exposure to 5 ppm were compared with those after sham exposure using a paired *t*-test or Wilcoxon's matched-pairs signed-rank test, as appropriate, and a significance of < 0.05. The D'Agostino and Pearson omnibus normality test was used to assess value distribution. Data are expressed as means ± SD or median with interquartile range (IQR, 25th;75th percentile). Data were analyzed and visualized by GraphPad Prism version 5.01 for Windows (GraphPad Software, San Diego, CA).

3 Results

3.1 Complaints and Clinical Symptoms

Complaints concerning eye, nose, and throat or airway discomfort were not reported during exposure to 0.05 ppm ethyl acrylate (sham). Also, no exposure-related effects were observed indicating irritation of the eye, and upper or lower respiratory tract. In the 5 ppm condition, olfactory sensations were predominantly labeled 'weak' and two subjects reported nausea. Complaints and signs of irritation concerning eye or nose were in the range 'weak' to 'moderate', whereas throat irritation rarely occurred. The lower airways were not affected by acute ethyl acrylate exposure. Detailed results are given in Fig. 1.

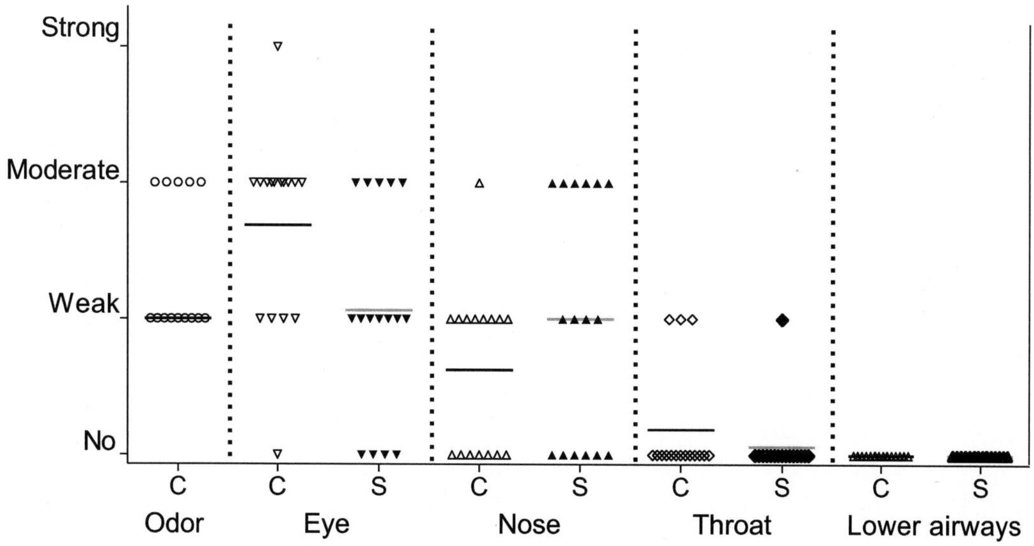

Fig. 1 Intensity of self-reported complaints (C) and observed symptoms (S) concerning relevant target sites after exposure to ethyl acrylate (5 ppm) for 4 h

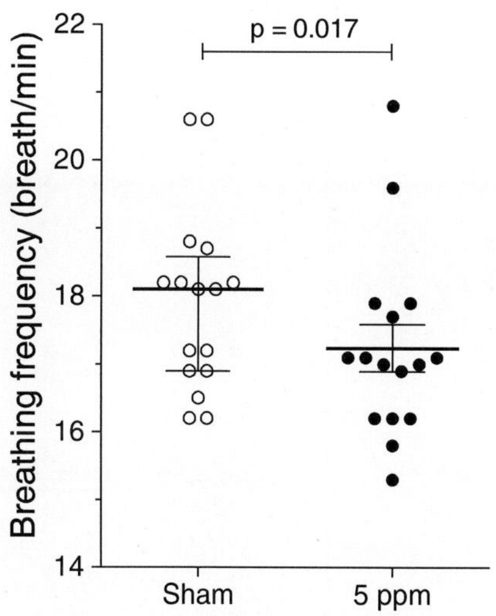

Fig. 2 Breathing frequency during 4 h of exposure to 0.05 ppm (sham) and 5 ppm ethyl acrylate

3.2 Breathing Frequency

The comparison of the median values of the respective 25 intervals on an individual basis revealed a significant difference between the two exposure conditions in 11 out of the 16 volunteers. The overall median breathing frequency during exposure was 18.1 (IQR 16.9; 18.6) for sham and 17.1 (IQR 16.2; 17.9) in case of ethyl acrylate ($p = 0.017$; Fig. 2). No influence of the order of the challenge conditions on respiratory rate could be assessed ($p = 0.296$, data not shown).

3.3 Biomarkers

The baseline level of biomarkers did not differ before sham challenge and 5 ppm ethyl acrylate. EBC-pH was 5.906 ± 0.087 and 5.917 ± 0.086, respectively; $p = 0.63$. FeNO value was 13.9 ± 5.3 ppb and 14.8 ± 5.1 ppb, respectively; $p = 0.199$. Significant changes of biomarkers could be observed after both exposure conditions. EBC-pH determined immediately after ethyl acrylate exposure increased significantly up to 5.996 ± 0.079; $p < 0.001$. After sham exposure we observed a similar pattern with pH values of 6.031 ± 0.062; $p < 0.0001$ (Fig. 3a). The detected increase in EBC-pH (ΔpH) after exposure was lower compared to sham exposure (ethyl acrylate 1.36 ± 1.28 % *vs.* sham 2.12 ± 1.09 %; $p = 0.036$) (Fig. 3b). The sham-adjusted effect of ethyl acrylate exposure on EBC-pH was a 0.76 % decrease after exposure. When comparing the first and following measurement

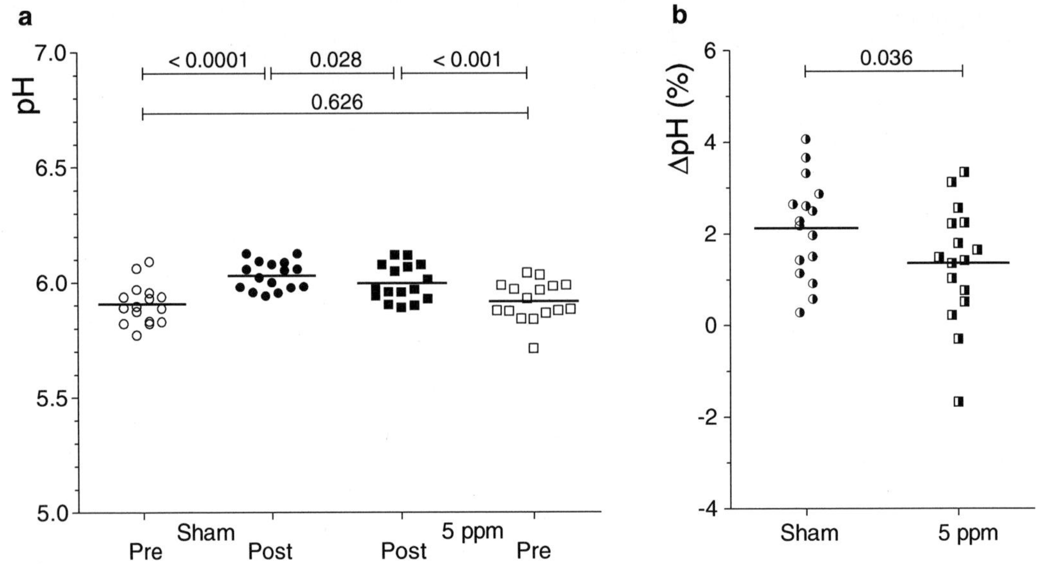

Fig. 3 Changes in pH after sham exposure (*circle*) and exposure to 5 ppm ethyl acrylate (*squares*) for 4 h; (**a**) EBC-pH was assessed before (*pre, open symbols*) and immediately after (*post, closed symbols*) exposure; (**b**) Percent change after exposure compared to start of exposure (ΔpH) for both conditions

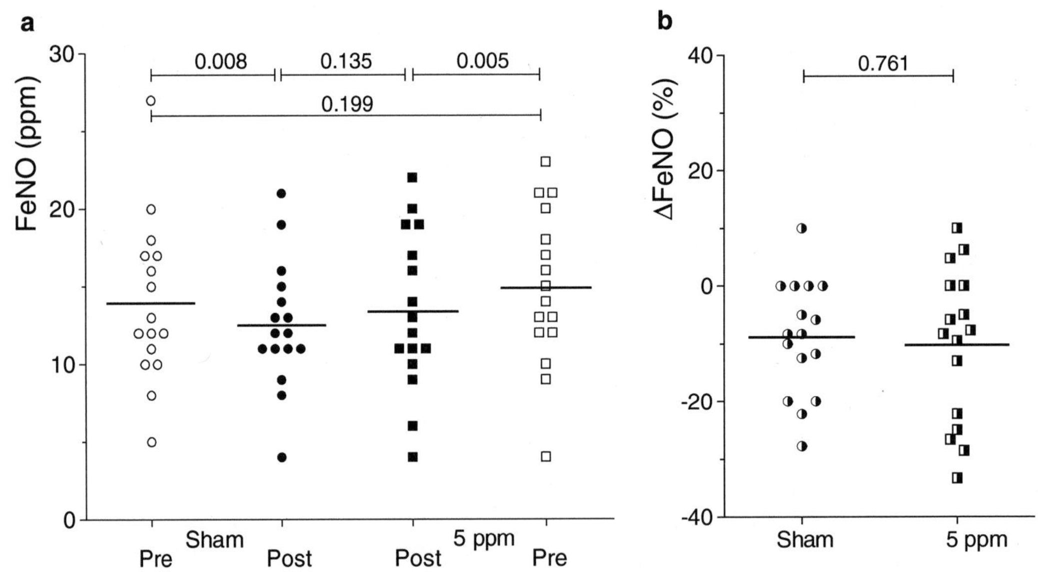

Fig. 4 Changes in FeNO after sham exposure (*circle*) and exposure to 5 ppm ethyl acrylate (*quadrat*) for 4 h; (**a**) FeNO was assessed before (*pre, open symbols*) and immediately after (*post, closed symbols*) exposure; (**b**) Percent change after exposure compared to start of exposure (ΔFeNO) for both conditions

independently from the exposure intensity, there was no significant difference ($p = 0.660$, data not shown).

In contrast to pH, FeNO was reduced after 4 h of challenge. We observed a significant decline after both sham (post 12.5 ± 4.1 ppb; $p = 0.008$) and ethyl acrylate exposure at a concentration of 5 ppm (post 13.4 ± 5.1 ppb; $p = 0.005$), respectively (Fig. 4a). Changes in FeNO levels (ΔFeNO) did not differ significantly after ethyl acrylate exposure compared to sham exposure (sham -8.9 ± 10.0 % *vs.* ethyl acrylate -10.3 ± 13.4 %; $p = 0.761$, Fig. 4b). Thus, no significant adjusted net-effect on FeNO could be observed after ethyl acrylate exposure. No influence of the order of the challenge conditions on FeNO could be identified ($p = 0.836$, data not shown).

3.4 Associations

Changes of FeNO were correlated with changes of EBC-pH after ethyl acrylate challenge ($r = 0.591$; $p = 0.016$). This was more apparent without adjusting for sham condition ($r = 0.753$; $p < 0.001$). The intensity of complaints and clinical signs of ocular or nasal irritation were not associated with changes in EBC-pH or FeNO after exposure to 5 ppm ethyl acrylate (data not shown). There was no association between reported complaints or observed signs of ocular irritation and changes in breathing frequency (data not shown). Concerning the nose no association could be revealed between clinical signs and breathing frequency, but in case of a reported nasal irritation a pronounced decrease in breathing rate could be observed (no *vs.* nasal irritation, 0 (−4.3; 4.1) *vs.* −5.5 (−10.0; −3.0) %, $p = 0.039$).

4 Discussion

Volatile agents can stimulate neural reflexes followed by triggering of a neurogenic inflammation (sensory irritation) or act directly through their physicochemical means and reaction products (tissue irritation) (Lee and Yu 2014). Different measurements of sensory irritation can be distinguished including behavioral, sensory, and physiological effects (Brüning et al. 2014; Arts et al. 2006). In this study, the intensity ratings of sensory perception and sensory-mediated reflexes were evaluated. While controlled exposure to 5 ppm ethyl acrylate for 4 h led to symptoms and signs of sensory irritation of eyes and nose, the lower respiratory tract was not affected in these measures. This is a reasonable result as the sensory irritation potency is correlated to chemical reactions with a receptor. About half of the total amount of inhaled ethyl acrylate is absorbed in each section of the airways. Thus, the target amount absorbed per unit area is much higher in the upper than in the lower airways (Alarie et al. 1998). Ethyl acrylate in a concentration of 5 ppm was demonstrated to be within the variation range of its lateralization threshold (van Thriel et al. 2006). This chemosensory property confines trigeminal stimulation and provides some information about “irritating concentrations”. Accordingly, challenge with ethyl acrylate at a concentration of 5 ppm provoked trigeminal perceptions in our volunteers. Besides nasal complaints, the burning, dry, and itching eyes were also among the most common symptoms. This is in accord with the reports indicating that eye irritation, perceived as stinging and burning, is an accompaniment to trigeminal-evoked intranasal perceptions (Hummel 2000). Overall, the intensity of symptoms in our subjects was rated as weak to moderate.

Sensory irritants can also reduce the breathing frequency and thereby the total amount of inhaled airborne chemicals (Shusterman 2003). This is caused by trigeminal nerve endings that evoke a burning sensation of the nasal passages and inhibit respiration from that site. In contrast, pulmonary irritants (like ozone) increase the respiratory rate accompanied by a decreasing tidal volume resulting in a rapid shallow breathing (Arts et al. 2006). The current challenge study was done under resting conditions characterized by nasal breathing in humans. In contrast to exercise with changing ratios between nasal and oral breathing, resting conditions ensure that nasal mucosa is the first contact site of inhaled ethyl acrylate within the respiratory tract. The observed reduction of the breathing frequency during 5 ppm ethyl acrylate exposure in our study went along with the perception of nasal irritation and could be interpreted as being in line with trigeminal stimulation. However, it is noteworthy that the reported “significance” refers to a statistical point of view and it is to question whether the assessed decrease in the range of breathing rate gains any significant clinical relevance.

In contrast to the eyes and upper airways, the lower airways' mucosal surface can only be inspected by invasive methods. Another approach for the evaluating of inflammatory responses is the analysis of biochemically changes. Biomarkers can be assessed non-invasively in exhaled breath (FeNO) and exhaled breath condensate (pH) (Hoffmeyer et al. 2009). Levels of these biomarkers can be altered by different cell types composing the mucous membranes, linking changes to tissue irritation. Also, effects starting along the sensory irritation pathway might encroach on non-sensory epithelial cells (Brüning et al. 2014). Our results were derived from healthy non-smoking subjects in order to exclude the known confounding of smoking on FeNO (Kharitonov et al. 1995) or EBC-pH (Koczulla et al. 2010). Moreover, with respect to possible circadian variations of biomarkers under study, sham and ethyl acrylate exposure at 5 ppm took place at the same time of day (Antosova et al. 2009). The determination of pH was performed in accordance with the published guidelines (Horváth et al. 2005) considering the influence of pCO_2. Therefore, EBC-pH and pCO_2 were simultaneously measured with a blood gas analyzer followed by calculation of $pH_{5.33}$ (Hoffmeyer et al. 2015a; Kullmann et al. 2007). Thus, our results seem reasonable even though pH changes across challenge were rather small. We observed a significant EBC-pH increase after the exposure to 5 ppm ethyl acrylate and also after sham exposure of a magnitude corresponding to that reported previously (Hoffmeyer et al. 2015b; Riediker and Danuser 2007). When adjusting for sham exposure, an overall negative net change of pH resulted. In view of these observations it is to be stressed that EBC-pH is an integrative measure of acids to bases ratio which can be affected by different mechanisms. It is well known that inflammation in several diseases is characterized by a pH decrease or acidification of the airways (Kostikas et al. 2002). Exposure to clean air itself might influence EBC-pH by the absence of otherwise common airborne acid contaminants. In our exposure unit, a pre-filter with the pore size F7, according to DIN ISO 2002, and a built-in HEPA-filter (High Efficiency Particulate Arrestor) realizes clean air requirements (Monsé et al. 2012). Constituents of inhaled air are capable to modulate the pH of EBC through physiochemical means. In this respect, ethyl acrylate is degraded mostly by hydrolysis to acrylic acid and ethanol. Exposure to alkaline products in cleaners (Corradi et al. 2012) and inhalation of an alkalinizing buffer (Davis et al. 2013) were reported to induce a pH increase of EBC. The pattern of FeNO changes in our study was inverted to pH changes demonstrating lower FeNO values after both challenge conditions. This result is in accord with the proposed interaction of airway acidity and nitrogen oxides. In this respect, airway acidification can trigger upregulation of nitric oxide synthase. In more detail, acid converts nitrite to nitric oxide through protonation to nitrous acid (Hunt 2006). As a consequence, less acid load could be linked to lower nitric oxide levels.

5 Conclusions

In summary, our results confirm the findings that ethyl acrylate is a sensory irritant for humans (Arts et al. 2006; Paustenbach 2000). Mucous membranes of eyes and upper airways are the prime target regions. Concerning the lower airways, an increase of EBC-pH after ethyl acrylate exposure was observed which was lower compared to sham condition. Whether this higher net acid load can be interpreted in terms of additional airway irritation, e.g., tissue irritation or already inflammatory alterations needs further investigations.

Acknowledgement We gratefully acknowledge the technicians of IPA Jennifer Gili, Anja Molkenthin, Melanie Ulbrich, and Susann Widmer.

Conflicts of Interest All the authors declare that they have no competing interests that might be perceived to influence the results and discussion reported in this manuscript.

References

Alarie Y, Schaper M, Nielsen GD, Abraham MH (1998) Structure-activity relationships of volatile organic chemicals as sensory irritants. Arch Toxicol 72:125–140

Antosova M, Bencova A, Psenkova A, Herle D, Rozborilova E (2009) Exhaled nitric oxide – circadian variations in healthy subjects. Eur J Med Res 14:6–8

Arts JH, de Heer C, Woutersen RA (2006) Local effects in the respiratory tract: relevance of subjectively measured irritation for setting occupational exposure limits. Int Arch Occup Environ Health 79:283–298

ATS/ERS (2005) ATS/ERS recommendations for standardized procedures for the online and offline measurement of exhaled lower respiratory nitric oxide and nasal nitric oxide. Am J Respir Crit Care Med 171:912–930

Blaszkewicz M, Hey K, Kiesswetter E, Kleinbeck S, Schäper M, van Thriel C (2010) Abgrenzung und Differenzierung "irritativer" und "belästigender" Effekte von Gefahrstoffen. Abschlussbericht. DGUV (Deutsche Gesetzliche Unfallversicherung); http://www.dguv.de/medien/ifa/de/pro/pro1/ff-fp0267/ff_fp0267_abschlussbericht.pdf. Accessed on 17 Dec 2015

Brüning T, Bartsch R, Bolt HM, Desel H, Drexler H, Gundert-Remy U, Hartwig A, Jäckh R, Leibold E, Pallapies D, Rettenmeier AW, Schlüter G, Stropp G, Sucker K, Triebig G, Westphal G, van Thriel C (2014) Sensory irritation as a basis for setting occupational exposure limits. Arch Toxicol 88:1855–1879

Cometto-Muñiz JE, Cain WS (1990) Thresholds for odor and nasal pungency. Physiol Behav 48:719–725

Corradi M, Gergelova P, Di Pilato E, Folesani G, Goldoni M, Andreoli R, Selis L, Mutti A (2012) Effect of exposure to detergents and other chemicals on biomarkers of pulmonary response in exhaled breath from hospital cleaners: a pilot study. Int Arch Occup Environ Health 85:389–396

Davis MD, Walsh BK, Dwyer ST, Combs C, Vehse N, Paget-Brown A, Pajewski T, Hunt JF (2013) Safety of an alkalinizing buffer designed for inhaled medications in humans. Respir Care 58:1226–1232

Hoffmeyer F, Raulf-Heimsoth M, Brüning T (2009) Exhaled breath condensate and airway inflammation. Curr Opin Allergy Clin Immunol 9:16–22

Hoffmeyer F, Berresheim H, Beine A, Sucker K, Brüning T, Bünger J (2015a) Methodological implications in pH standardization of exhaled breath condensate. J Breath Res 9(3):036003

Hoffmeyer F, Sucker K, Monsé C, Berresheim H, Jettkant B, Rosenkranz N, Brüning T, Bünger J (2015b) Different patterns in changes of exhaled breath condensate ph and exhaled nitric oxide after ozone exposure. Adv Exp Med Biol 834:39–47

Horváth I, Hunt J, Barnes PJ, Alving K, Antczak A, Baraldi E, Becher G, van Beurden WJ, Corradi M, Dekhuijzen R, Dweik RA, Dwyer T, Effros R, Erzurum S, Gaston B, Gessner C, Greening A, Ho LP, Hohlfeld J. Jöbsis Q, Laskowski D, Loukides S, Marlin D, Montuschi P, Olin AC, Redington AE, Reinhold P, van Rensen EL, Rubinstein I, Silkoff P, Toren K, Vass G, Vogelberg C, Wirtz H; ATS/ERS Task Force on Exhaled Breath Condensate (2005) Exhaled breath condensate: methodological recommendations and unresolved questions. Eur Respir J 26:523–548

Hummel T (2000) Assessment of intranasal trigeminal function. Int J Psychophysiol 36:147–155

Hunt J (2006) Airway acidification: interactions with nitrogen oxides and airway inflammation. Curr Allergy Asthma Rep 6:47–52

Kharitonov SA, Robbins RA, Yates D, Keatings V, Barnes PJ (1995) Acute and chronic effects of cigarette smoking on exhaled nitric oxide. Am J Respir Crit Care Med 152:609–612

Kobal G, Van Toller S, Hummel T (1989) Is there directional smelling? Experientia 45:130–132

Koczulla AR, Noeske S, Herr C, Jörres RA, Römmelt H, Vogelmeier C, Bals R (2010) Acute and chronic effects of smoking on inflammation markers in exhaled breath condensate in current smokers. Respiration 79:61–67

Kostikas K, Papatheodorou G, Ganas K, Psathakis K, Panagou P, Loukides S (2002) pH in expired breath condensate of patients with inflammatory airway diseases. Am J Respir Crit Care Med 165:1364–1370

Kullmann T, Barta I, Lázár Z, Szili B, Barát E, Valyon M, Kollai M, Horváth I (2007) Exhaled breath condensate pH standardised for CO_2 partial pressure. Eur Respir J 29:496–501

Lee LY, Yu J (2014) Sensory nerves in lung and airways. Comp Physiol 4:287–324

MAK (2012) The MAK collection for occupational health and safety, Wiley Online Library, Online ISBN: 9783527600410. doi:10.1002/3527600418

Miller RR, Young JT, Kociba RJ, Keyes DG, Bodner KM, Calhoun LL, Ayres JA (1985) Chronic toxicity and oncogenicity bioassay of inhaled ethyl acrylate in Fischer 344 rats and B6C3F1 mice. Drug Chem Toxicol 8:1–42

Monsé C, Sucker K, van Thriel C, Broding HC, Jettkant B, Berresheim H, Wiethege T, Käfferlein H, Merget R, Bünger J, Brüning T (2012) Considerations for the design and technical setup of a human whole-body exposure chamber. Inhal Toxicol 24:99–108

Paustenbach DJ (2000) The history and biological basis of occupational exposure limits for chemical agents. In:

Harris R (ed) Patty's industrial hygiene, vol 3, 5th edn. Wiley, New York, pp 1903–2000
Quanjer PH, Stanojevic S, Cole TJ, Baur X, Hall GL, Culver BH, Enright PL, Hankinson JL, Ip MS, Zheng J, Stocks J (2012) Multi-ethnic reference values for spirometry for the 3-95-yr age range: the global lung function 2012 equations. Eur Respir J Suppl 40:1324–1343
Quirce S, Lemière C, de Blay F, del Pozo V, Gerth Van Wijk R, Maestrelli P, Pauli G, Pignatti P, Raulf-Heimsoth M, Sastre J, Storaas T, Moscato G (2010) Noninvasive methods for assessment of airway inflammation in occupational settings. Allergy 65:445–458
Raulf-Heimsoth M, Hoffmeyer F, van Thriel C, Blaszkewicz M, Bünger J, Brüning T (2010) Assessment of low dose effects of acute sulphur dioxide exposure on the airways using non-invasive methods. Arch Toxicol 84:121–127
Riediker M, Danuser B (2007) Exhaled breath condensate pH is increased after moderate exercise. J Aerosol Med 20:13–18
Shusterman D (2003) Toxicology of nasal irritants. Curr Allergy Asthma Rep 3:258–265
Smith MG, Croy I, Ögren M, Waye KP (2013) On the influence of freight trains on humans: a laboratory investigation of the impact of nocturnal low frequency vibration and noise on sleep and heart rate. PLoS One 8(2):e55829. doi:10.1371/journal.pone.0055829
van Thriel C, Schäper M, Kiesswetter E, Kleinbeck S, Juran S, Blaszkewicz M, Fricke HH, Altmann L, Berresheim H, Brüning T (2006) From chemosensory thresholds to whole body exposures-experimental approaches evaluating chemosensory effects of chemicals. Int Arch Occup Environ Health 79:308–321

Advs Exp. Medicine, Biology - Neuroscience and Respiration (2016) 24: 21–26
DOI 10.1007/5584_2016_246

Published online: 10 May 2016

Effectiveness of PCR and Immunofluorescence Techniques for Detecting Human Cytomegalovirus in Blood and Bronchoalveolar Lavage Fluid

A. Roży, K. Duk, B. Szumna, P. Skrońska, D. Gawryluk, and J. Chorostowska-Wynimko

Abstract

Current diagnostic methods allow a rapid and reliable detection of active human cytomegalovirus (hCMV) infection by identifying the presence of pp65 CMV antigen or CMV DNA in peripheral blood and affected organs. The goal of this study was to evaluate the effectiveness of CMV detection in blood and organ-specific biological material, such as bronchoalveolar lavage fluid (BALF), by comparing two standard diagnostic methods, immunofluorescence (IF) and the real-time polymerase chain reaction (PCR). We evaluated 25 patients with concomitant respiratory disease who were referred to our hospital for diagnosis due to suspected acute CMV infection. The presence of hCMV was concomitantly evaluated by IF and PCR in 16 peripheral blood samples. In two patients, we observed positive results for both IF and PCR, and in two other patients the results were discordant. Of 11 patients, CMV DNA was detected in six BALF samples, and in one blood plasma sample. Real-time PCR detected CMV DNA in 54.6 % of BALF samples and 12.0 % of blood samples, while indirect IF testing confirmed antigenemia in 12.5 % of blood samples. The results from our study suggest that the IF method is as effective as PCR for detecting an ongoing CMV infection in blood samples. However, real-time PCR was much more effective at detecting CMV DNA in BALF compared to blood samples. Our results suggest that the biological material being tested during CMV diagnosis should be derived directly from the virally infected organ(s).

A. Roży, K. Duk, B. Szumna, P. Skrońska, and J. Chorostowska-Wynimko (✉)
Department of Genetics and Clinical Immunology, National Institute of Tuberculosis and Lung Diseases, 26 Płocka St, 01-138 Warsaw, Poland
e-mail: j.chorostowska@igichp.edu.pl

D. Gawryluk
Third Department of Lung Disease, National Institute of Tuberculosis and Lung Diseases, Warsaw, Poland

Keywords
Blood • Bronchoalveolar lavage fluid • Human cytomegalovirus • Lungs • pp65 antigen • Viral infection

1 Introduction

Human cytomegalovirus (hCMV; human herpesvirus 5; HH5) is a common pathogen worldwide. The prevalence of antibodies against CMV ranges from 40 % in Europe and 70 % in the US to greater than 80 % in developing countries. The widespread presence of CMV is associated with its high infectious potential and effective routes of transmission. CMV is generally transmitted through direct contact with body fluids from infected individuals, but also through transplanted organs, bone marrow, and blood transfusions that contain viral particles.

CMV can infect a variety of human cells including parenchymal, connective tissue, and hematopoietic cells. The predominant cells targeted for viral replication are epithelial, endothelial, fibroblasts, and smooth muscle. Similar to other herpesviruses, CMV persists in latency, typically in phagocytic cells, for the entire life of the host. This mechanism plays an important role in the spread of infection throughout the body (Sinzger et al. 2008). In immunocompetent individuals, CMV infections are typically asymptomatic although persistent infections can occur. In contrast, in immunodeficient patients, i.e. patients with cancer, systemic diseases, and prescribed immune suppressive therapies, CMV infections can trigger severe complications (Poole et al. 2014).

Current diagnostic methods allow for rapid and reliable confirmation of hCMV. Serological evaluation, i.e. assessment of serum levels of anti-CMV IgG and/or IgM antibodies, which assesses the viral dynamics and avidity, is also an easy and available diagnostic tool. However, the practicality of this method is restricted and it is usually used retrospectively to confirm already diagnosed cases. More valuable methods allow for the direct confirmation of the presence of CMV in samples. CMV identification using conventional cell culture methods is effective due to its excellent specificity and sensitivity. This method uses biological samples inoculated onto human fibroblast cells, incubated and assessed for cytopathic effects, e.g. viral shedding. However, due to very specific laboratory requirements and the long processing time (2–3 weeks to confirm a negative result), cell culture is not routinely used for CMV diagnostics. CMV propagation in fibroblast culture followed by viral antigen detection by indirect immunofluorescence (IF) may be a useful alternative as it allows for rapid viral detection after 16 h of incubation (Jahan 2010; Ross et al. 2011). The highest diagnostic value is currently attributed to the direct detection of specific pp65 CMV antigen, a major structural late protein expressed in blood leukocytes during the early phase of the CMV replication cycle. Alternatively, the gene encoding pp65 can be used to quantitatively and qualitatively assess the presence of CMV DNA (Chevillotte et al. 2009). Quantitative evaluation of viral DNA copies is a useful tool for both viral detection and monitoring of the infection, in terms of its intensity and duration. Importantly, the real-time polymerase chain reaction (PCR) assay is sensitive enough to detect hCMV in body fluids other than blood and urine (Jahan 2010).

This study presents a preliminary comparison of the effectiveness of IF compared to PCR for the detection of pp65 antigenemia for CMV diagnosis in patients with acute or exacerbated chronic inflammatory processes in the lungs. Both methods allow for the specific detection of an active CMV infection prior to the onset of clinical symptoms.

2 Methods

2.1 Patients

The study protocol was accepted by a local Ethics Committee of the National Institute of Tuberculosis and Lung Diseases in Warsaw,

Poland. Twenty five patients (17–78 years of age, 12 women and 13 men) with concomitant respiratory disease (interstitial lung disease n = 8, autoimmune disease n = 11, transplanted organ n = 2, cancer n = 2, respiratory failure n = 1, and pneumonia n = 1) referred for the diagnostics due to suspected acute CMV infection were included in the study. In 11 patients paired, simultaneously collected materials from peripheral blood and lower respiratory tract (bronchoalveolar lavage fluid, BALF) were obtained for analyses. In 14 subject, only peripheral blood was collected. All tests were performed as a part of routine laboratory diagnostics.

2.2 hCMV pp65 Antigen Detection

The human CMV antigenemia was analyzed within 2 h of specimen collection using the standard indirect immunofluorescence CMV Brite™ Turbo kit (IQ Products BV; Groningen, Netherlands) in accordance with the manufacturer's instructions. Briefly, the cytospin slides, with 200,000 cells per glass slide, were prepared, fixed and permeabilized. The presence of CMV pp65 antigen was detected by using a cocktail of two monoclonal antibodies (C10/C11) directed against CMV lower matrix phosphoprotein (pp65) and visualized with a fluorescent (FITC) secondary antibody. The results were expressed as the number of positive cells per slide. The test was considered positive when ≥1 fluorescent cell was observed for every 200,000 leukocytes in fluorescence microscope at 400x magnification.

2.3 hCMV DNA detection

The CMV DNA identification was carried out alongside the CMV antigenemia assay. The hCMV DNA was identified by ready-to-use CMV R-gene™ Kit (Argene; Verniolle, France) according to the manufacturer's instructions. It detects and measures the CMV genome after viral DNA extraction. It works by detecting and simultaneously amplifying a specific region of the CMV DNA – UL83 gene encoding pp65 tegument protein, using 5' nuclease Taqman technology. Size of amplified fragment is 283 base pairs.

Briefly, CMV DNA was isolated from 200 μL of blood or BALF samples using the QIAamp® DNA Blood Mini kit (Argene; Verniolle, France). The PCR reaction was performed according to manufacturer's protocol on Light Cycler 480 II (Roche; Basel, Switzerland). $C_t < 40$ was accepted as the laboratory-determined limit of detection.

3 Results

3.1 Immunofluorescence (IF) *vs.* Real-Time PCR for Human Cytomegalovirus (hCMV) Detection in Blood Samples

The presence of hCMV was concomitantly evaluated by IF and PCR in 14 peripheral blood samples. A double positive result was obtained for one sample by the PCR reaction ($C_t = 35.0$) in addition to a positive result for pp65, as shown by indirect IF (9 pp65 positive cells/slide). For 11 patients, both specimens provided a negative result, and no CMV DNA or pp65 antigen were detected. In one sample, a positive result was observed only by PCR ($C_t = 34.7$). In addition, one sample was negative according to the PCR results and inconclusive by the IF assessment, which was due to the nonspecific cytoplasmatic staining on the blood cytospin slides (Table 1).

Table 1 hCMV detection by indirect immunofluorescence (IF) *vs.* real-time PCR in peripheral blood samples from subjects with suspected acute CMV infection and concomitant respiratory disease

	Blood	
Result	IF	PCR
Positive	1/14 (7.1 %)	2/14 (14.3 %)
Negative	12/14 (85.7 %)	12/14 (85.7 %)
Inconclusive	1/14 (7.1 %)	0/14 (0.0 %)

Table 2 hCMV detection by real-time PCR in paired peripheral blood and bronchoalveolar lavage fluid (BALF) samples collected from subjects with suspected acute CMV infection and concomitant respiratory disease

	Blood	BALF
Result	PCR	
Positive	0/9 (0 %)	4/9 (44.4 %)
Negative	9/9 (100 %)	4/9 (44.4 %)
Inconclusive	0/9 (0 %)	1/9 (11.1 %)

3.2 Real-Time PCR for CMV Detection in Peripheral Blood *vs.* Bronchoalveolar Lavage Fluid (BALF)

Paired peripheral blood and BALF samples from nine patients were analyzed using real-time PCR. CMV DNA was detected in four BALF (44.4 %) samples, but was not observed in any of the respective blood samples (9/9) (Table 2).

3.3 IF *vs.* Real-Time PCR for CMV Detection in Blood *vs.* Bronchoalveolar Lavage Fluid (BALF)

Two sets of paired samples from blood and BALF were simultaneously analyzed for antigenemia and hCMV DNA. In one patient, presence of CMV virus was confirmed in blood by IF and in both blood and BALF by real-time PCR (365 pp65^{+}cells/slide; $C_t = 26.0$ for blood *vs.* $C_t = 23.7$ for BALF). In the second patient, the PCR test identified CMV DNA in BALF ($C_t = 27.4$), but the blood sample was inconclusive ($C_t = 40$). The concomitant indirect IF assessment was not reliable due to nonspecific staining (Table 3).

In total, real-time PCR evaluation detected CMV DNA in 54.6 % of BALF samples (6/11) and in 12.0 % of blood samples (3/25). Indirect IF testing confirmed the presence of pp65 in 12.5 % of the blood samples (2/16). Positive and negative controls were always performed in parallel for IF and real-time PCR assessment.

Table 3 hCMV detection by indirect immunofluorescence (IF) *vs.* real-time PCR in paired peripheral blood and bronchoalveolar lavage fluid (BALF) samples collected from subjects with suspected acute CMV infection and concomitant respiratory disease

	Blood		BALF
	IF	PCR	PCR
Patient 1	+	+	+
Patient 2	+/−	+/−	+

(+) – positive, (+/−) – inconclusive

4 Discussion

This study demonstrates the effectiveness of two most common laboratory methods used for rapid diagnosis of acute CMV infection: detection of a specific hCMV antigen, pp65, by IF and real-time PCR to detect viral genetic material. Both are commonly used for monitoring viral infection and tracking recurrence, and are useful when considering CMV therapy initiation. pp65, a CMV lower matrix protein, is also an early antigen in viral replication and is abundantly present in antigen-positive polymorphonuclear cells (Jahan 2010). A major limitation associated with this method is that peripheral blood is the only sample type that can be accurately evaluated. Alternatively, PCR-based diagnostic tests are applicable to any biological material containing CMV DNA, due to their high sensitivity. However, high sensitivity also increases the risk of false positives in contaminated samples. False negative results may also be due to the presence of PCR inhibitors in the sample.

In the present study we found that IF was positive for hCMV in 12.5 % (2/16) of peripheral blood samples. PCR detected hCMV DNA in 18.8 % (3/16) of blood samples. These findings are in line with other studies on the subject. Kwon et al. (2015), analyzing blood from transplant recipients for CMV using IF and real-time PCR, have reported that pp65 antigenemia assay is capable of detecting the presence of CMV in 20.7 % (99/479) of the samples, while real-time PCR detects hCMV DNA in 32.6 % (156/479).

Likewsie, Zipeto et al. (1992) have detected the pp65 with indirect IF in 42.3 % (124/293) of blood samples obtained from immunocompromised patients, and real-time PCR detected hCMV DNA in 62.5 % (183/293) of the samples. Both methods similarly detected hCMV in blood samples. However, the slightly more frequent detection of DNA using PCR may have been due to CMV being in the latent state.

We further investigated the practical utility of BALF samples as a material originating directly from diseased organs. We found that real-time PCR testing detected hCMV DNA in 54.6 % (6/11) of BALF samples and 9.1 % (1/11) of concomitant blood samples. Bauer et al. (2007) have reported a significantly higher levels of hCMV DNA in the lung compared with the blood from lung transplant recipients. In that study, PCR was positive for hCMV DNA in 52.6 % (51/97) and 26.8 % (26/97) of paired BALF and blood samples, respectively. Likewise, CMV DNA is more frequently detected by PCR in BALF (42.3 % [11/26]) than in blood samples (38.5 % [10/26]) from asymptomatic bone marrow recipients (Fajac et al. 1997). Such results have also been observed for HIV patients with pulmonary symptoms (52.9 % [81/153] for BALF samples *vs.* 39.5 % [49/124] for serum samples) (Hansen et al. 1997).

Differences in detection rates between methods may be attributed to ongoing acute respiratory CMV infection, and to the fact that the lungs are the reservoir for CMV. Dworniczak et al. (2004) have reported significantly higher hCMV DNA copy numbers in BALF than in blood leukocytes in patients with idiopathic pulmonary fibrosis, and in healthy controls (log10 = 2.7 *vs.* 1.2 for patients and 2.8 *vs.* 0.9 for controls). Interestingly, there was no difference between the relative number of cases positive for hCMV DNA between the patients (75 %) and controls (69 %), corroborating the notion of hCMV infection being widespread.

The findings of the present study show that PCR is a highly effective and sensitive tool for detecting hCMV DNA. The discrepancy between the blood and BALF results may be due to a high sensitivity of PCR, which enables the detection of small quantities of genetic material.

It should be emphasized that detection of hCMV DNA does not necessarily indicate an active CMV infection. Thus, results of a diagnostic test should always be interpreted in the context of the overall clinical symptoms (Jones 2014). A quantitative assessment of hCMV DNA copies is particularly helpful for a precise, targeted decision to initiate antiviral therapy because it enables a continuous monitoring of viral copy numbers in the blood. In contrast, pp65 blood antigenemia enables the early detection of CMV infection in the absence of any clinical signs (Zipeto et al. 1992). Thus, the appropriate diagnostic method will ultimately depend on the patient's clinical symptoms.

Acknowledgments This study was performed as part of the scientific project (1/21) of the National Institute of Tuberculosis and Lung Diseases in Warsaw, Poland.

Conflicts of Interest The authors declare no conflicts of interest in relation to this article.

References

Bauer CC, Jaksch P, Aberle SW, Haber H, Lang G, Klepetko W, Hofmann H, Puchhammer-Stöckl E (2007) Relationship between cytomegalovirus DNA load in epithelial lining fluid and plasma of lung transplant recipients and analysis of coinfection with Epstein-Barr virus and human herpesvirus 6 in the lung compartment. J Clin Microbiol 45 (2):324–328

Chevillotte M, Landwehr S, Linta L, Frascaroli G, Lüske A, Buser C, Mertens T, von Einem J (2009) Major tegument protein pp 65 of human cytomegalovirus is required for the incorporation of pUL69 and pUL97 into the virus particle and for viral growth in macrophages. J Virol 83(6):2480–2490

Dworniczak S, Ziora D, Kapral M, Mazurek U, Niepsuj G, Rauer R, Wilczok T, Kozielski J (2004) Human cytomegalovirus DNA level in patients with idiopathic pulmonary fibrosis. J Physiol Pharmacol 55 (Suppl 3):67–75

Fajac A, Stéphan F, Ibrahim A, Gautier E, Bernaudin JF, Pico JL (1997) Value of cytomegalovirus detection by PCR in bronchoalveolar lavage routinely performed in asymptomatic bone marrow recipients. Bone Marrow Transplant 20(7):581–585

Hansen KK, Vestbo J, Benfield T, Lundgren JD, Mathiesen LR (1997) Rapid detection of cytomegalovirus in bronchoalveolar lavage fluid and serum samples by polymerase chain reaction: correlation of virus isolation and clinical outcome for patients with

human immunodeficiency virus infection. Clin Infect Dis 24(5):878–883
Jahan M (2010) Laboratory diagnosis of CMV infection: a review. Bangladesh J Med Microbiol 04(02):39–44
Jones RP (2014) A study of an unexplained and large increase in respiratory deaths in England and Wales: is the pattern of diagnoses consistent with the potential involvement of cytomegalovirus? Br J Med Med Res 4 (33):5193–5217
Kwon S, Jung BK, Ko SY, Lee CK, Cho Y (2015) Comparison of quantitation of cytomegalovirus DNA by real-time PCR in whole blood with the cytomegalovirus antigenemia assay. Ann Lab Med 35:99–104
Poole E, Wills M, Sinclair J (2014) Human cytomegalovirus latency: targeting differences in the latently infected cell with a view to clearing latent infection. N J Sci 2014:10. doi:10.1155/2014/313761
Ross SA, Novak Z, Pati S, Boppana SB (2011) Diagnosis of cytomegalovirus infections. Infect Disord Drug Targets 11(5):466–474
Sinzger C, Digel M, Jahn G (2008) Cytomegalovirus cell tropism. Curr Top Microbiol Immunol 325:63–83
Zipeto D, Revello MG, Silini E, Parea M, Percivalle E, Zavattoni M, Milanesi G, Gerna G (1992) Development and clinical significance of a diagnostic assay based on the polymerase chain reaction for detection of human cytomegalovirus DNA in blood samples from immunocompromised patients. J Clin Microbiol 30(2):527–530

Advs Exp. Medicine, Biology - Neuroscience and Respiration (2016) 24: 27–35
DOI 10.1007/5584_2016_247

Published online: 02 July 2016

The Role of Ion Channels to Regulate Airway Ciliary Beat Frequency During Allergic Inflammation

M. Joskova, M. Sutovska, P. Durdik, D. Koniar, L. Hargas, P. Banovcin, M. Hrianka, V. Khazaei, L. Pappova, and S. Franova

Abstract

Overproduction of mucus is a hallmark of asthma. The aim of this study was to identify potentially effective therapies for removing excess mucus. The role of voltage-gated (Kir 6.1, K_{Ca} 1.1) and store-operated ion channels (SOC, CRAC) in respiratory cilia, relating to the tracheal ciliary beat frequency (CBF), was compared under the physiological and allergic airway conditions. *Ex vivo* experiments were designed to test the local effects of Kir 6.1, K_{Ca} 1.1 and CRAC ion channel modulators in a concentration-dependent manner on the CBF. Cilia, obtained with the brushing method, were monitored by a high-speed video camera and analyzed with ciliary analysis software. In natural conditions, a Kir 6.1 opener accelerated CBF, while CRAC blocker slowed it in a concentration-dependent manner. In allergic inflammation, the effect of Kir 6.1 opener was insignificant, with a tendency to decrease CBF. A cilio-inhibitory effect of a CRAC blocker, while gently reduced by allergic inflammation, remained significant. A K_{Ca} 1.1 opener turned out to significantly enhance the CBF under the allergic OVA-sensitized conditions. We conclude that optimally attuned concentration of K_{Ca} 1.1 openers or special types of bimodal SOC channel blockers, potentially given by inhalation, might benefit asthma.

M. Joskova (✉), M. Sutovska, V. Khazaei, L. Pappova, and S. Franova
Department of Pharmacology, BioMed Martin, Jessenius Faculty of Medicine, Comenius University in Bratislava, 4C Mala Hora St, 036 01 Martin, Slovakia
e-mail: joskova@jfmed.uniba.sk

P. Durdik and P. Banovcin
Department of Children and Adolescents, Jessenius Faculty of Medicine, Comenius University in Bratislava and Martin University Hospital, 2 Kollarova St, 036 01 Martin, Slovakia

D. Koniar, L. Hargas, and M. Hrianka
Department of Mechatronics and Electronics, Faculty of Electrical Engineering, University of Zilina, 1 Univerzitna St, 010 26 Zilina, Slovakia

Keywords
Asthma • Ciliary beat frequency • Mucociliary transport • Store-operated ion channels • Voltage-gated ion channels

1 Introduction

Ion channels are transmembrane proteins through which ions pass according to their electrochemical gradient. They are gated by voltage, second messengers, and other intracellular/extracellular mediators and are implicated in a multitude of pathophysiological processes, including respiratory system diseases (Valverde et al. 2011).

Out of the large spectrum of ion channels listed in the IUPHAR nomenclature, we focused our attention on voltage-gated and store-operated ion channels (SOC), particularly on the inwardly rectifying potassium channels (Kir 6.1), the calcium-activated potassium channels (potassium large conductance calcium-activated channels K_{Ca} 1.1), and the calcium release-activated calcium ion channels (CRAC).

In the airways, potassium channels contribute to the bronchodilator responses and control neuronal reflexes, the production of mucus and its secretion from goblet cells, the reduction of microvascular permeability, and the modulation of mucociliary clearance and epithelial cell restoration (Manzanares et al. 2011, 2014; Sutovska et al. 2013). All beneficial features of the potassium channel openers could be advantageous in the therapy of chronic pulmonary diseases if they are bronchoselective. In pulmonary medicine, inhalation is the preferred route of administration with the lower risk of systemic side effects. Only have a few experimental studies been devoted to the relationship of inhaled potassium channel openers and bronchodilation (Kidney et al. 1996). Cumulative single doses of potassium channel openers have been studied in adult patients with mild-to-moderate non-allergic asthma, but without confirmation of their bronchodilator efficacy (Faurschou et al. 1994). Purkey et al. (2014) have investigated the association between chronic rhinosinusitis and human genetic variants in two airway epithelial potassium channels (K_{Ca} 1.1 and voltage-dependent Kv 7.5). K^+ channel genes have been confirmed in a greater number in Paramecium, a genus of ciliated protozoan, a representative of the ciliate group, than in humans (Haynes et al. 2003).

Many medications can alter the potassium channel function. Hypoglycemic agents derived from sulfonylurea, vasodilatory drug used to treat angina, nicorandil, and antiarrhythmic and inotropic (levosimendan) agents are known as the channel modulators. The ion channel expression differs under pathophysiological conditions. Asthma is associated with the loss of K_{Ca} 1.1 channel function and the upregulation of sensor (STIM1) and structural (Orai1) protein components of CRAC ion channels (Spinelli et al. 2012).

These latter ion channels respond to the depletion of endoplasmic reticulum (ER) calcium stores as a consequence of inositol triphosphate signal transduction, followed by the store-operated calcium entry from extracellular space *via* Orai1. This plasma membrane channel, through coupling with translocated STIM sensor protein, replenishes ER calcium stores. CRAC blocker regulate changes in defense mechanisms of the airways, e.g., cough reflex and bronchoconstriction under the asthma experimental conditions (Sutovska et al. 2013). They are also involved in secretory functions of mast cells, T cells, and eosinophils (Di Capite et al. 2011).

The inflammatory cells mentioned above and the structural cells of the airways, including epithelial cells, are major sources of mediator-driven chronic inflammation in asthma, the pathological features of which include bronchospasm, plasma exudation, mucus secretion, airway hyperreactivity and structural changes (Barnes 2003). Mucus overproduction makes expectoration more difficult. Therefore, stimulation of ciliary beating may increase and support mucociliary transport to prevent airway obstruction by viscous mucus plugs. Several experimental studies have documented changes in the ion channel expression and function of the airway epithelium (Galietta et al. 2004; Anagnostopoulou et al. 2010). Therefore, we hypothesized that the modulation of ion channels might affect the motor component of mucociliary clearance as well. We addressed the issue by examining the function of the Kir 6.1, K_{Ca} 1.1. and CRAC ion channels in the airway ciliary movement in the physiological condition and experimental allergen-induced airway inflammation.

2 Methods

The study protocol was approved by a local Ethics Committee of the Jessenius Faculty of Medicine in Martin, Slovakia. The experiments were in accord with the EU criteria for experimental animal welfare (EK 996/2012).

Experiments were performed in adult Trik strain male guinea pigs, weighing 250–300 g after a minimum 1-week quarantine period. The animals were obtained from the Department of Experimental Pharmacology, Slovak Academy of Sciences, Dobra Voda, Slovakia, and they were kept in an animal house with the recommended temperature, humidity, ventilation rate, noise levels and 12:12 h day/night cycles. The animals were housed in groups of maximum four *per* cage, with ready access to fresh water and a proper diet.

2.1 Study Design

The guinea pigs were randomly divided into the following experimental groups, consisting of eight animals each:

- Control – healthy, treated with 0.9 % NaCl for 21 days
- OVA – sensitized to ovalbumin allergen, treated with 0.9 % NaCl for 21 days

The following ion channel modulators were used:

- Pinacidil – Kir 6.1 channel opener
- Glibenclamide – Kir 6.1 channel blocker
- NS1619 – K_{Ca} 1.1 channel opener
- TEA – K_{Ca} 1.1 channel blocker
- FPCA – CRAC channel blocker

The groups with administration of substances to the tracheal cilia of healthy animals:

- P – pinacidil 10^{-7}, 10^{-6}, and 10^{-5} mol.l^{-1}
- P + G – pinacidil 10^{-7}, 10^{-6}, and 10^{-5} mol.l^{-1} added to glibenclamide 10^{-6}, 10^{-5}, and 10^{-4} mol.l^{-1}
- NS – NS1619 10^{-7}, 10^{-6}, and 10^{-5} mol.l^{-1}
- NS + TEA – NS1619 10^{-7}, 10^{-6}, and 10^{-5} mol.l^{-1} added to TEA 10^{-6}, 10^{-5}, and 10^{-4} mol.l^{-1}
- FPCA – FPCA 10^{-7}, 10^{-6}, and 10^{-5} mol.l^{-1}
- DMSO – 10 % DMSO
- SB – positive control – salbutamol 10^{-4} mol.l^{-1}

The groups with administration of substances to the tracheal cilia of allergen-sensitized animals:

- PS – pinacidil 10^{-7}, 10^{-6}, and 10^{-5} mol.l^{-1} (PS7, PS6, PS5)
- NSS – NS1619 10^{-7}, 10^{-6}, and 10^{-5} mol.l^{-1} (NSS7, NSS6, NSS5)
- FPCAS – FPCA 10^{-7}, 10^{-6}, and 10^{-5} mol.l^{-1} (FPCAS7, FPCAS6, FPCAS5)
- DMSOS – 10 % DMSO
- SBS – positive control – salbutamol 10^{-4} mol.l^{-1}

2.2 Chemicals

The following chemical agents were purchased from Sigma-Aldrich (Hamburg, Germany): DMSO – dimethyl sulfoxide, pinacidil monohydrate – (±)-N-cyano-N′-4-pyridinyl-N″-(1,2,2-trimethylpropyl) guanidine monohydrate, glibenclamide – 5-chloro-N-[4-(cyclohexylureidosulfonyl) phenethyl]-2-methoxybenzamide, glyburide – N-p-[2-(5-chloro-2-methoxybenzamido) ethyl] benzenesulfonyl-N′-cyclohexylurea, NS1619 – 1,3-dihydro-1-[2-hydroxy-5-(trifluoromethyl) phenyl]-5-(trifluor omethyl)-2H-benzimidazol-2-one, TEA – tetraet hylammonium chloride. FPCA – 3-fluoropyridine -4-carboxylic acid was purchased from Alfa Aesar (Karlsruhe, Germany) and RPMI 1640 medium from Invitrogen/Life Technologies Gibco (Waltham, MA). Pinacidil, glibenclamide, and NS1619, were dissolved in 10 % DMSO before they were diluted to a definite concentration of 10^{-5} mol.l^{-1}, 10^{-6} mol.l^{-1}, and 10^{-7} mol.l^{-1}. TEA was dissolved in 0.9 % NaCl to a concentration of 10^{-5} mol.l^{-1}, 10^{-6} mol.l^{-1}, and 10^{-7} mol.l^{-1}. FPCA was dissolved in water for injection prior to its dilution in the saline to a concentration of 10^{-5} mol.l^{-1}, 10^{-6} mol.l^{-1}, and 10^{-7} mol.l^{-1}.

2.3 Ovalbumin-Induced Allergic Inflammation of Airways

The ovalbumin allergen causes airway reactivity changes *via* immunological mechanisms. Aluminium hydroxide, $Al(OH)_3$, adjuvant is known as a Th_2 inducer. A suspension of ovalbumin of 10^{-5} mol.l^{-1} in $Al(OH)_3$ was administered over a period of 21 days. Guinea pigs received concurrent 0.5 ml OVA intraperitoneal and subcutaneous injections on Day 1 of sensitization, 0.5 ml OVA intraperitoneally alone on Day 4, and 0.5 ml OVA subcutaneously on Day 14. The degree of sensitization was confirmed by responses to allergen (1 % OVA), given by inhalation, once a day for 1–3 min on Days 15–21 through a PARI Jet Nebulizer (Paul Ritzau, Pari-Werk GmbH; Starnberg, Germany; output 5 l.s^{-1}, particle mass median diameter 1.2 μm) attached to a double-chamber whole body plethysmograph (HSE type 855, Hugo Sachs Elektronik, Germany). All animal experiments were initiated 1 week after the last allergen exposure.

2.4 Ciliary Beat Frequency Analysis

An analysis of ciliary beat frequency was carried out in an *in vitro* laboratory air-conditioned setting, with controlled temperature (21–24 °C) and relative humidity (approximately 55 ± 10 %). Temperature of the cilia RPMI 1640 medium (ThermoFisher Scientific; Waltham, MA) and the microscopic glass slide was maintained in a range of 37–38 °C by a PeCon 2000–2 Temp Controller (PeCon GmbH; Erbach, Germany). After sacrificing the animals, a transverse access to the anterior tracheal wall was made by the midline incision of neck tissues. Ciliated samples were obtained by brushing the tracheal wall, with a cytology brush of 2.5 mm in diameter. The brush was dipped into the saline, gently rotated on the mucosal surface of the trachea and then removed. Cilia were suspended in 1 ml of RPMI 1640 Medium and used to make a microscopic preparation. Only were undisrupted beating ciliated cells recorded with a digital high-speed video camera (Basler A504kc; Basler AG, Germany) at a frame rate of 256–512 *per* sec. The camera was connected with both inverted phase contrast microscope (Zeiss Axio Vert. A1; Carl Zeiss AG, Göttingen, Germany) and a computer. There were approximately 10–12 sequential image recordings, each approximately 10 s in duration, of the same preparation performed at 1 min intervals.

Video records were analyzed using ciliary analysis software (LabVIEW™) to generate the ciliary region of interest (ROI), the intensity variation in selected ROIs, and the intensity variance curve. The curve was subjected to the fast Fourier transform (FFT) algorithm. The Fourier spectrum of each intensity variance curve was then equal to the frequency spectrum of beating in selected ROIs. ROIs were finally compared with the relevant video record to filter out artefacts.

2.5 Statistical Analysis

The median of ciliary beat frequency for each ROI and the arithmetic mean of a set of ROI values for each sample were used to determine the ciliary beat frequency (CBF) expressed in Hertz (Hz). All data were expressed as means ± SE. Statistical significance of differences was assessed with one-way ANOVA with *post-hoc* Bonferroni test. A p-value of <0.05 was used to define significant differences.

3 Results

In healthy guinea pigs, the opener of Kir 6.1 potassium ion channels pinacidil (10^{-5} mol.l^{-1}, 10^{-6} mol.l^{-1}), caused a significant increase in CBF (*p<0.05). This effect was concentration-dependent and was abolished in the presence of the non-specific Kir 6.1 blocker glibenclamide (10^{-4} mol.l^{-1} and 10^{-5} mol.l^{-1}) (Fig. 1a). In contrast, during airway allergic inflammation, the opening of Kir 6.1 channels by pinacidil (10^{-7}, 10^{-6}, and 10^{-5} mol.l^{-1}) was associated with a tendency to a decrease in CBF (Fig. 1b).

The opener of K_{Ca} 1.1 channels NS1619 (10^{-7}, 10^{-6}, and 10^{-5} mol.l^{-1}) had no effect on the CBF in the physiological condition (Fig. 2a), but its highest concentration (10^{-5} mol.l^{-1}) caused a significant enhancement of CBF in the airway inflammatory condition (Fig. 2b).

The selective blocker of SOC (CRAC) ion channels FPCA (10^{-7} mol.l^{-1}, 10^{-6} mol.l^{-1}, 10^{-5} mol.l^{-1}) significantly reduced the ciliary movement, almost in a concentration-dependent manner (*p<0.05, **p<0.01, ***p<0.001, respectively) in the physiological condition (Fig. 3a). This decrease was mitigated but still significant by FPCA concentration of 10^{-5} mol.l^{-1} in allergic inflammatory conditions (Fig. 3b).

4 Discussion

In this study we demonstrate the role of potassium (Kir 6.1 and K_{Ca} 1.1) and calcium (CRAC) ion channels in the regulation of tracheal ciliary beat frequency (CBF) in healthy and ovalbumin-sensitized guinea pigs. We confirmed the crucial role of Kir 6.1 and CRAC ion channels in the modulation of the CBF in both experimental conditions. Kir 6.1 channels were engaged in the tracheal ciliostimulation in the healthy condition, which is in line with the findings of Ohba et al. (2013), who have demonstrated a relationship between pharmacological K_{ATP} stimulation and acceleration of ciliary movement in mice. These authors show that activation of

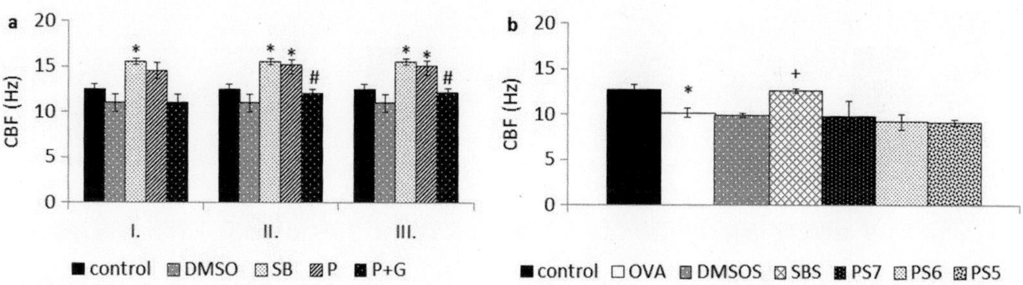

Fig. 1 The role of Kir 6.1 ion channels in the regulation of ciliary beat frequency (CBF) in unsensitized and ovalbumin (OVA)-sensitized animals after local application of DMSO, salbutamol, and pinacidil and glibenclamid in *in vitro* condition. Tracheal cilia were exposed to the agents always after brushing. (**a**) physiological conditions: I – pinacidil (10^{-7} mol.l^{-1}), glibenclamid (10^{-6} mol.l^{-1}); II – pinacidil (10^{-6} mol.l^{-1}), glibenclamid (10^{-5} mol.l^{-1}); and III – pinacidil (10^{-5} mol.l^{-1}), glibenclamid (10^{-4} mol.l^{-1}). Control group – cilia of healthy guinea pigs exposed to saline; DMSO – cilia of healthy guinea pigs exposed to 10 % DMSO; SB – cilia of healthy guinea pigs exposed to salbutamol (10^{-4} mol.l^{-1}); P – cilia of healthy guinea pigs exposed to pinacidil (10^{-7} mol.l^{-1}; 10^{-6} mol.l^{-1}; 10^{-5} mol.l^{-1}), an opener of Kir 6.1; P + G – cilia of healthy guinea pigs exposed to pinacidil (10^{-7} mol.l^{-1}; 10^{-6} mol.l^{-1}; 10^{-5} mol.l^{-1}) plus glibenclamid (10^{-6} mol.l^{-1}; 10^{-5} mol.l^{-1}; 10^{-4} mol.l^{-1}); (**b**) allergic condition consisting of OVA-sensitized guinea pigs treated for 21 days with 0.9 % NaCl – control bar is same as in Panel A; DMSOS – cilia of OVA-sensitized group exposed to 10 % DMSO; SBS – cilia of OVA-sensitized group exposed to salbutamol (10^{-4} mol.l^{-1}); PS7 – cilia of OVA-sensitized group exposed to pinacidil (10^{-7} mol.l^{-1}), an opener of Kir 6.1; PS6 – cilia of OVA-sensitized group exposed to pinacidil (10^{-6} mol.l^{-1}); PS5 – cilia of OVA-sensitized group exposed to pinacidil (10^{-5} mol.l^{-1}) (Data are expressed as means ± SE; n = 8; *$p < 0.05$ compared with the control group; $^{+}p<0.05$ compared with the OVA group; $^{\#}p < 0.05$ compared with the P group)

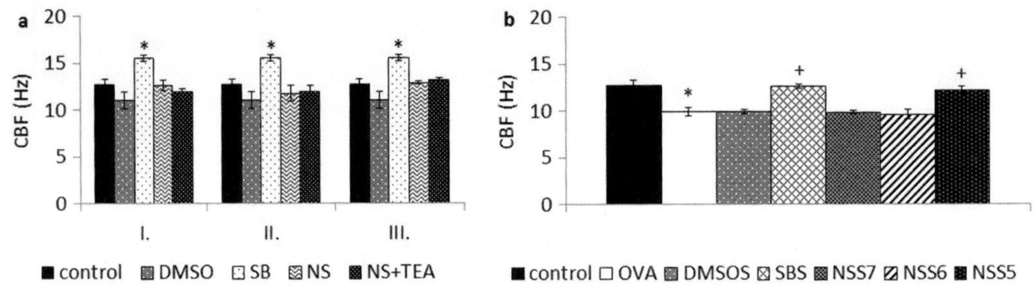

Fig. 2 The role of K_{Ca} 1.1 ion channels in the regulation of ciliary beat frequency (CBF) in unsensitized and ovalbumin (OVA)-sensitized guinea pigs after local application of DMSO, salbutamol, and NS1619 and tetraethylammonium chloride (TEA) in *in vitro* condition. Tracheal cilia were exposed to the agents always after brushing. (**a**) physiological condition: I – NS1619 (10^{-7} mol.l^{-1}), TEA (10^{-6} mol.l^{-1}); II – NS1619 (10^{-6} mol.l^{-1}), TEA (10^{-5} mol.l^{-1}); and III – NS1619 (10^{-5} mol.l^{-1}), TEA (10^{-4} mol.l^{-1}); Control group – cilia of healthy guinea pigs exposed to saline; DMSO – cilia of healthy guinea pigs exposed to 10 % DMSO; SB – cilia of healthy guinea pigs exposed to salbutamol (10^{-4} mol.l^{-1}); NS – cilia of healthy guinea pigs exposed to NS1619 (10^{-7} mol.l^{-1}; 10^{-6} mol.l^{-1}; 10^{-5} mol.l^{-1}), an opener of K_{Ca} 1.1; NS + TEA – cilia of healthy guinea pigs exposed to NS1619 (10^{-7} mol.l^{-1}; 10^{-6} mol.l^{-1}; 10^{-5} mol.l^{-1}) plus TEA (10^{-6} mol.l^{-1}; 10^{-5} mol.l^{-1}; 10^{-4} mol.l^{-1}), a blocker of K_{Ca} 1.1; (**b**) allergic condition consisting of OVA-sensitized guinea pigs treated for 21 days with 0.9 % NaCl – control bar is same as in Panel A; DMSO – cilia of OVA-sensitized group exposed to 10 % DMSO; SBS – cilia of OVA-sensitized group exposed to salbutamol (10^{-4} mol.l^{-1}); NSS7 – cilia of OVA-sensitized group exposed to the NS1619 (10^{-7} mol.l^{-1}), an opener of K_{Ca} 1.1; NSS6 – cilia of OVA-sensitized group exposed to NS1619 (10^{-6} mol.l^{-1}); NSS5 – cilia of OVA -sensitized group exposed to NS1619 (10^{-5} mol.l^{-1}) (Data are expressed as means ± SE; n = 8; *$p < 0.05$ compared with the control group; $^{+}p<0.05$ compared with the OVA group)

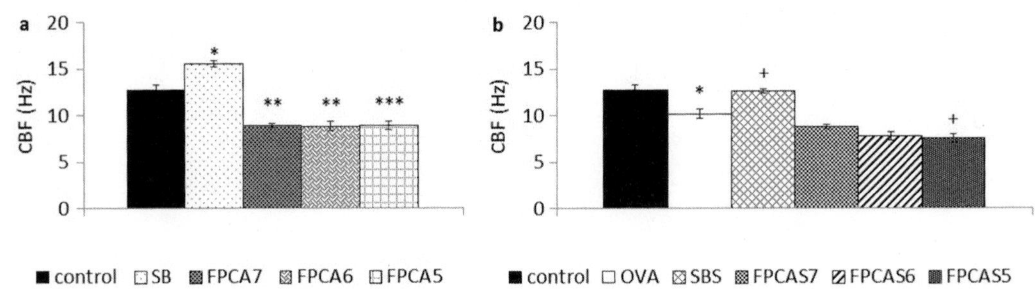

Fig. 3 The role of SOC (CRAC) ion channels in the regulation of ciliary beat frequency (CBF) in unsensitized and ovalbumin (OVA)-sensitized animals after local application of 3-fluoropyridine-4-carboxylic acid (FPCA) in *in vitro* condition. Tracheal cilia were exposed to the agents always after brushing. (**a**) physiological conditions: Control group – cilia of healthy guinea pigs exposed to saline; SB – cilia of healthy guinea pigs exposed to salbutamol (10^{-4} mol.l^{-1}); FPCA7/FPCA6/FPCA5 – cilia of healthy guinea pigs exposed to FPCA (10^{-7} mol.l^{-1}; 10^{-6} mol.l^{-1}; 10^{-5} mol.l^{-1}), a blocker of CRAC; (**b**) allergic condition consisting of OVA-sensitized guinea pigs treated for 21 days with 0.9 % NaCl – control bar is same as in Panel A; SBS – cilia of OVA-sensitized group exposed to salbutamol (10^{-4} mol.l^{-1}); FPCAS7/FPCAS6/FPCAS5 – cilia of OVA-sensitized group exposed to FPCA (10^{-7} mol.l^{-1}; 10^{-6} mol.l^{-1}; 10^{-5} mol.l^{-1}), a blocker of CRAC (Data are expressed as means ± SE; n = 8; *$p < 0.05$, **$p<0.01$, and ***$p<0.001$ compared with the control group; $^{+}p<0.05$ compared with the OVA group)

non-voltage dependent calcium channels (non-VDCC) is a consequence of membrane hyperpolarization induced by a K_{ATP} opener. We demonstrate in the present study that ciliary movement slows down when Kir 6.1 ion channels become open in allergic inflammation. That, in turn, seems in line with an argument that CBF decreases with increasing external mucus viscosity until it reaches a plateau (Liedtke and Heller 2007).

There have only been a few reports on the identification of potassium ion channels in ciliates (Haynes et al. 2003). Past research has established an association between calcium-activated potassium channels and swimming behavior in cilia (Valentine et al. 2012). In contrast, our present results demonstrate that large-conductance K_{Ca} 1.1 channels were not crucial in affecting the tracheal CBF in healthy guinea pigs, but became important in pathology. These channels are abundant in normal smooth muscle cells of airways, where they regulate membrane potentials and the process of muscle contraction. In human bronchial epithelium, apical K_{Ca} 1.1 channels regulate surface liquid volume (Manzanares et al. 2011). However, these channels have not been identified in the mouse tracheal epithelium (Schreiber et al. 2002). It is of interest that cytokines, typical for allergic asthma, may have diverse effects on of K_{Ca} 1.1 channel activity. Whereas interleukin-4 (IL-4) provides a stimulatory input, IL-13 partly antagonizes the effect of IL-4 (Martin et al. 2008). A human study of Laoukili et al. (2001) reported a time- and dose-dependent inhibitory effect of the Th_2 cytokine, IL-13, on the nasal CBF. Different sensitivities of individual subfamilies of calcium-activated potassium channels to inhibitors have also been determined. The sequence homology of transmembrane hydrophobic cores has revealed the following differences: BK channels are large conductance K_{Ca} 1.1 channels inhibited by TEA, SK channels are small conductance K_{Ca} 2.1, 2.2, and 2.3 channels, IK channels are intermediate conductance K_{Ca} 3.1 channels, and the other subfamilies – K_{Ca} 4.1, 4.2, and K_{Ca} 5.1 are structurally related to K_{Ca} 1.1 channels, but insensitive to internal Ca^{2+} (Perez-Zoghbi et al. 2009). It is possible that other subtypes of potassium channels, such as K_{Ca} 3.1, could be involved in the alteration of CBF during natural conditions.

In the present study we also demonstrate the importance of SOC ion channels of airway cilia in the control of CBF in the healthy condition. A possible explanation of this role may be a deficit in the calcium replenishment of empty stores in the endoplasmic reticulum resulting from SOC ion channel inhibition. Calcium ions belong to the intracellular signals that mediate changes in CBF in response to different stimuli. SOC ion channels have also been implicated in the regulation of brain ependymal cilia (Nguyen et al. 2001). In the present study, inhibitory effect on ciliary beating of SOC antagonism persisted, although mitigated, in ovalbumin-driven allergic inflammation. The SOC channels could participate in maintaining ciliary movement during inflammation as a result of action of inflammatory mediators. Prostaglandins and histamine released during allergic inflammation influence mucociliary clearance, acting alone or in combination with other mediators. Prostaglandin E1 (PGE1) increases tracheal CBF in the guinea pig and enhances the stimulatory effect of histamine on CBF in the rabbit maxillary sinus. Nonetheless, histamine does not appreciably influence CBF in the guinea pig, due likely to interspecies differences in responsiveness to inflammatory mediators (Khan et al. 1995; Dolata 1990).

CRAC channels in airway cilia play a notable role in the pathophysiology of allergic airway inflammation (Di Capite et al. 2011). The tracheal surface in guinea pigs is abundant in ciliated cells, which makes these animals much suitable for airway epithelium studies (Li et al. 2011). The concentration of the CRAC ion channel blocker used in the present study was identical to that used in other experiments that confirmed the blocker's ability to weaken the cough reflex and airway resistance (Sutovska et al. 2013). Therefore, allergic inflammation might modify the expression and function of CRAC channels.

Respiratory infections often evoked by ciliary dysfunction might reduce the amount of intact cilia and, along with hypersecretion of mucus, they can intensify a vicious inflammatory cycle. We thus submit that modulation of CBF by potassium (Kir 6.1, K_{Ca} 1.1) and SOC ion channels could plausibly be an expression of airway defense mechanisms. Medicines with potential ciliostimulating effects might act beneficially by aiding the mucociliary defense mechanism. As CRAC channels exhibit bimodal concentration-dependent responses to their blockade (Jairaman and Prakriya 2013), nebulization of specific CRAC blocking agents, yet to be designed, could benefit the management of airway inflammatory conditions. On the other side, Kir 6.1 channel openers, known as potential bronchodilators, do not appear to have a positive effect on CBF in asthma. By contrast, K_{Ca} 1.1 openers could ameliorate a disturbed natural cleaning function of airway cilia in inflammation, providing these agents are given directly to the airways, which would also help avoid systemic side effects. A better understanding of the ion channels'pecularities would be of therapeutic potential in the pathological states characterized by deranged function of cilia in the tracheal epithelial cells in airway allergic inflammatory conditions.

Acknowledgement This research was conducted within the project 'Measurement of Respiratory Epithelium Cilium Kinematics' and was supported by the grants VEGA 1/0165/14; MZ 2012/35-UK MA-12; APVV-0305-12; BioMed Martin (ITMS 26220220187); Center of Experimental and Clinical Respirology II, and 'The increasing of opportunities for career growth in research and development in the medical sciences', co-financed from EU sources. The authors would like to thank Katarina Jesenska for technical help during the experiments.

Conflict of Interests All authors declared no conflicts of interest in relation to this article.

References

Anagnostopoulou P, Dai L, Schatterny J, Hirtz S, Duerr J, Mall MA (2010) Allergic airway inflammation induces a pro-secretory epithelial ion transport phenotype in mice. Eur Respir J 36:1436–1447

Barnes PJ (2003) Pathophysiology of asthma. European Respiratory Monograph 23:84–113

Di Capite JL, Bates GJ, Parekh AB (2011) Mast cell CRAC channel as a novel therapeutic target in allergy. Curr Opin Allergy Clin Immunol 11:33–38

Dolata J (1990) Prostaglandin E1 enhances the histamine induced stimulation of the mucociliary activity in the rabbit maxillary sinus. Eur Respir J 3:559–565

Faurschou P, Mikkelsen KL, Steffensen I, Franke B (1994) The lack of bronchodilator effect and the short-term safety of cumulative single doses of an inhaled potassium channel opener (bimakalim) in adult patients with mild to moderate bronchial asthma. Pulmonary Pharmacology Therapeutics 7:293–297

Galietta LJV, Folli C, Caci E, Pedemonte N, Taddei A, Ravazzolo R, Zegarra-Moran O (2004) Effect of inflammatory stimuli on airway ion transport. Proc Am Thorac Soc 1:62–65

Haynes WJ, Ling KY, Saimi Y, Kung C (2003) PAK paradox: paramecium appears to have more K^+-channel genes than humans. Eukaryotic Cell 2:737–745

Jairaman A, Prakriya M (2013) Molecular pharmacology of store-operated CRAC channels. Channels 7:402–414

Khan R, Dolata J, Lindberg S (1995) Effects of inflammatory mediators on ciliary function in vitro. Rhinology 33:22–25

Kidney JC, Lotvall JO, Lei Y, Chung KF, Barnes PJ (1996) The effect of inhaled K+ channel openers on bronchoconstriction and airway microvascular leakage in anaesthetised guinea pigs. Eur J Pharmacol 296:81–87

Laoukili J, Perret E, Willems T, Minty A, Parthoens E, Houcine O, Coste A, Jorissen M, Marano F, Caput D, Tournier F (2001) IL-13 alters mucociliary differentiation and ciliary beating of human respiratory epithelial cells. J Clin Investig 108:1817–1824

Li Y, Wang J, He HY, Ma LJ, Zeng J, Deng GC, Liu X, Engelhardt JF, Wang Y (2011) Immunohistochemical demonstration of airway epithelial cell markers of guinea pig. Tissue Cell 43:283–290

Liedtke WB, Heller S (eds) (2007) The TRP ion channel function in sensory transduction and cellular signaling cascades. The TRPV4 channel in ciliated epithelia. CRC Press Taylor and Francis Group, USA, Boca Raton, Chapter 30. Available from: http://www.ncbi.nlm.nih.gov/books/NBK5252/. Accessed on 20 Jan 2016

Manzanares D, Gonzalez C, Ivonnet P, Chen RS, Valencia-Gattas M, Conner GE, Larsson HP, Salathe M (2011) Functional apical large conductance, Ca^{2+}-activated, and voltage-dependent K+ channels are required for maintenance of airway surface liquid volume. J Biol Chem 286:19830–19839

Manzanares D, Srinivasan M, Salathe ST, Ivonnet P, Baumlin N, Dennis JS, Conner GE, Salthe M (2014) IFN-γ-mediated reduction of large-conductance, Ca2 + –activated, voltage-dependent K+ (BK) channel activity in airway epithelial cells leads to mucociliary dysfunction. Am J Physiol Lung Cell Mol Physiol 306:L453–L462

Martin G, O'Connell RJ, Pietrzykowski AZ, Treistman SN, Ethier MF, Madison JM (2008) Interleukin-4 activates large-conductance, calcium activated potassium (BKCa) channels in human airway smooth muscle cells. Exp Physiol 93:908–918

Nguyen T, Chin WC, O'Brien JA, Verdugo P, Berger AJ (2001) Intracellular pathways regulating ciliary beating of rat brain ependymal cells. J Physiol 531 (Pt 1):131–140

Ohba T, Sawada E, Suzuki Y, Yamamura H, Ohya S, Tsuda H, Imaizumi Y (2013) Enhancement of Ca2+ influx and ciliary beating by membrane

hyperpolarization due to ATP-sensitive K^+ channel opening in mouse airway epithelial cells. J Pharmacol Exp Ther 347:145–153
Perez-Zoghbi JF, Karner C, Ito S, Shepherd M, Alrashdan YA, Sanderson MJ (2009) Ion channel regulation of intracellular calcium and airway smooth muscle function. Pulmonary Pharmacology Therapeutics 22:388–397
Purkey MT, Li J, Mentch F, Grant SFA, Desrosiers M, Hakonarson H, Toskala E (2014) Genetic variation in genes encoding airway epithelial potassium channels is associated with chronic rhinosinusitis in a pediatric population. PLoS ONE 9:e89329. doi:10.1371/journal.pone.0089329
Schreiber R, Mürle B, Sun J, Kunzelmann K (2002) Electrolyte transport in the mouse trachea: no evidence for a contribution of luminal K^+ conductance. J Membr Biol 189:143–151
Spinelli AM, González-Cobos JC, Zhang X, Motiani RK, Rowan S, Zhang W, Garrett J, Vincent PA, Matrougui K, Singer HA, Trebak M (2012) Airway smooth muscle STIM1 and Orai1 are upregulated in asthmatic mice and mediate PDGF-activated SOCE, CRAC currents, proliferation, and migration. Pflugers Arch 464:481–492
Sutovska M, Adamkov M, Kocmalova M, Mesarosova L, Oravec M, Franova S (2013) CRAC ion channels and airway defense reflexes in experimental allergic inflammation. Adv Exp Med Biol 756:39–48
Valentine MS, Rajendran A, Yano J, Weerarante SD, Beisson J, Cohen J, Koll F, van Houten J (2012) Paramecium BBS genes are key to presence of channels in Cilia. Cilia 1:16
Valverde MA, Cantero-Recasens G, Garcia-Elias A, Jung C, Carreras-Sureda A, Vicente R (2011) Ion channels in asthma. J Biol Chem 286:32877–32882

Advs Exp. Medicine, Biology - Neuroscience and Respiration (2016) 24: 37–44
DOI 10.1007/5584_2016_227

Published online: 31 May 2016

Content of Asthmagen Natural Rubber Latex Allergens in Commercial Disposable Gloves

C. Bittner, M.V. Garrido, L.H. Krach, and V. Harth

Abstract

The use of natural rubber latex (NRL) gloves in many occupations may lead to latex sensitization, allergic asthma, and skin reactions. Due to their good properties and environmental safety NRL gloves are still being used in the healthcare setting, but also in the food industry, by hairdressers, cleaners, etc. The aim of our study was to assess the protein and NRL allergen content in commercial gloves by different methods, including a new assay. Twenty commercially available NRL gloves were analyzed. Protein extraction was performed according to the international standard ASTM D-5712. Total protein content was measured with a modified Lowry method, NRL content with the CAP Inhibition Assay, the Beezhold ELISA Inhibition Assay, and an innovative ELISA with IgY-antibodies extracted from eggs of NRL-immunized hens (IgY Inhibition Assay). We found a high protein content in a range of 215.0–1304.7 μg/g in 8 out of the 20 NRL gloves. Seven of the 20 gloves were powdered, four of them with a high protein content. In gloves with high protein content, the immunological tests detected congruently high levels of NRL allergen. We conclude that a high percentage of commercially available NRL gloves still represent a risk for NRL allergy, including asthma. The modified Lowry Method allows to infer on the latex allergen content.

Keywords

IgY-antibodies • IgY inhibition assay • Latex allergy • Latex gloves

1 Introduction

Natural rubber latex (NRL) gloves have been in use since the nineteenth century, initially for protecting patients and healthcare workers

C. Bittner (✉), M.V. Garrido, L.H. Krach, and V. Harth
Division of Clinical Occupational Medicine, Institute for Occupational and Maritime Medicine (ZfAM), University Medical Center Hamburg-Eppendorf, 10 Seewartenstraße, 20459 Hamburg, Germany
e-mail: cordula.bittner@bgv.hamburg.de

(HCW) from communicable diseases. Over the last 30 years, the use of latex gloves has increased exponentially and its use has spread to other sectors, like the food industry, hairdressing, cleaning services, etc. (Mota and Turrini 2012). Although the allergenicity of NRL was recognized early, international attention to the relevancy of type I allergic reactions to latex was first developed from the beginning of the 1980s as skin problems and respiratory type I allergies rose to a major problem especially for HCW (Cabanes et al. 2012). At the time at which NRL with high allergen content and powdered NRL gloves were still in use, the prevalence of latex allergy among HCW was considerably high (up to 11 %) (Vandenplas 1995; Turjanmaa et al. 1988).

NRL is a cytoplasmic exudate of the lactic chyle of the tropical *Hevea brasiliensis* tree and contains most of the known proteins found in any plant cell (Lamberti et al. 2015). The high protein content bears a high allergenic potential of NRL materials. To date, up to 15 allergens and additional isoallergens have been included in the IUIS nomenclature (International Union of Immunological Societies Allergen Nomenclature Subcommittee 2015). One of the main reasons for the high risk of developing respiratory allergies was the powder layer used in the NRL gloves, since the powder acted as a carrier of the NRL allergens facilitating its inhalation (Tarlo et al. 1994). The protein content of NRL gloves and therefore their allergenic potential can be considerably reduced by eliminating the powder and through manufacturing procedures, like leaching (extensive washing processes). Thus, in order to protect HCW from NRL-related allergies, the use of powdered NRL gloves was banned for the healthcare sector in the 1990s. In addition, international regulations (i.e., European DIN standard EN 455-3) have recommended to limit the total protein content in NRL gloves for use in healthcare to the minimum possible. A guide value of a maximum of 30 μg protein per gram of glove material has been recommended in the corresponding German technical standard (BAuA 2008). The powdered gloves ban and the allergen contents limit are restricted to HCW. Regulations to protect workers in other sectors do not yet exist. Thus, occupations, such as grocers, hairdressers, and cleaners as well as users in non-occupational settings may still remain at relevant risk for developing latex allergy.

In addition, standards for the measurement of latex content in NRL gloves have been introduced (e.g., US-standard ASTM D 6499). The methods recommended in the standards have several disadvantages regarding standardization, reliability, and costs. Moreover, the Beezhold ELISA Inhibition Assay, which is considered the standard for the determination of latex allergen content in NRL products, raises major ethical concerns, since it requires to sacrifice the animals, which makes it difficult to reconcile with the current animal protection awareness.

In the present study we investigated the protein content and NRL allergen content of commercially available gloves. We compared different methods and developed a new test method with the aim to overcome the disadvantages of hitherto existing assays.

2 Methods

The study was approved, according to the German Law on Animal Protection, by the Hamburg Health Authority. An approval from the Ethics Committee was waived since no humans were involved in the study.

2.1 Latex Protein Extraction from NRL Gloves

A sample of 20 commercially available NRL gloves was drawn randomly from several German drugstores, supermarkets, pharmacies, and hospital material providers. NRL was extracted according to the method recommended in the US-Standard ASTM D 5712. Briefly, gloves weight was determined using an electronic balance. The gloves were cut into 1 × 1 cm pieces. The pieces were soaked in a glass container with

1 ml phosphate-buffered saline (PBS) per g glove material. The closed container was stirred during 2 h in a 37 °C water bath. The eluate was transferred to centrifugation tubes and centrifuged over 20 min at 4000 rpm. The aliquot was stored at −20 °C until protein or allergen analysis. In order to avoid any contamination, latex-free gloves were worn during the whole process.

2.2 Protein Quantification With a Modified Lowry Method

We quantified the total protein content with the Lowry Micro DC Protein Assay Kit (BioRad Laboratories; Munich, Germany) according to the manufacturer's instruction and the ASTM standard D 5712. Briefly, samples and standards were incubated in 1.5 ml Eppendorf-tubes stepwise with trichloroacetic acid (72 % TCA-solution), phosphor wolfram acid (72 % PTA-solution), and sodium deoxycholate (0.15 % DOC-solution) during 10–20 min at room temperature. The samples were centrifuged during 15 min at 6000 g. The supernatant was discarded and the pellet was solved in 250 μl 0.1 N sodium hydroxide. After 15 min incubation with 100 μl Reagent A at room temperature, Reagent B was added. The photometrical measurement was done in a disposable cuvette at 750 nm.

2.3 CAP Inhibition Assay

The latex allergen content was determined indirectly by IgE determination in the ImmunoCAP-system (FisherScientific; Freiburg, Germany) following IgE-inhibition of glove extracts as described earlier (Baur et al. 1997). Briefly, a standard dilution of Malaysian latex milk in concentrations from 0.005 to 4.0 μg/ml, on the one hand, and a pool of serum from 5 latex allergic volunteers, on the other hand, were incubated at 4 °C over night with the extracts of the 20 different latex gloves each. Subsequently, unbound NRL antibodies in the supernatant were measured by ImmunoCAP with NRL (k82; FisherScientific). The content of allergenic protein was determined from the standard curve of defined NRL content.

2.4 Beezhold ELISA Inhibition Assay

Two rabbits were vaccinated with 500 μl NRL-standard-antigen (1 mg/ml) and 500 μl complete Freund's adjuvant and 4 weeks later with incomplete Freund's adjuvant. The rabbits were exsanguinated 6 weeks after the first vaccination by transthoracic heart puncture and the serum was obtained. The inhibition assay was done according to the ASTM standard D 6499. Briefly, an inhibition step was performed as described above using the serum pool of the 2 immunized rabbits instead of the human serum pool. Unbound antibodies were measured by ELISA with anti-rabbit-IgG antibodies. As described in the standard, repeated determination was performed.

2.5 IgY Inhibition Assay

Two chicken were vaccinated with 500 μl NRL-standard-antigen solution (1 mg/ml) and 500 μl complete and incomplete Freund's adjuvant in an interval of 10 days. The eggs laid 2 weeks or later after vaccination were collected and stored at 4 °C. IgY antibodies were obtained from the egg yolk with the EGGSTract® IgY Purification System (Promega; Mannheim, Germany). The ELISA IgY-inhibition assay was built according to the above mentioned ASTM Standard D 6499, whereas IgY-antibodies were used instead of IgG-antibodies. The second antibody was an anti-Chicken-IgY conjugated-antibody (Promega; Mannheim, Germany).

2.6 Statistical Elaboration

The results of the tests were compared with Spearman's rank correlation coefficient (r_s). The accuracy of the allergen determination methods was assessed using the receiver operating characteristic (ROC) curve. The results of the modified Lowry Method were taken as reference to calculate sensitivity and specificity. A degree of agreement between tests was evaluated calculating kappa. All calculations were made with the statistical software package PASW Statistics 18.0.0 (IBM Germany).

3 Results

The results of the protein and allergen content analysis with the modified Lowry Method, the Beezhold ELISA Inhibition Assay, the CAP Inhibition Assay, and the IgY Inhibition Assay are presented in Table 1. The total protein content exceeded the recommended threshold for medical use gloves of 30 μg per gram glove material in 8 out of the 20 gloves, ranging from 215.0 to 1304.7 μg per gram glove material. Powdered gloves exceeded the recommended threshold more frequently than powder-free gloves (57 % *vs.* 31 %).

The results from the NRL allergen test methods correlated well with the results of the modified Lowry Method which measures total protein content (Table 2). The highest correlation with the total protein content was seen for the results of the Beezhold ELISA Inhibition Assay ($r_s = 0.883$). Our IgY Inhibition Assay showed a lower, but still satisfactory, correlation ($r_s = 0735$). The results of the IgY Inhibition Assay and of the CAP Inhibition Assay showed also a high correlation with the results from the Beezhold ELISA Inhibition Assay, which is considered to be the gold standard for determining latex allergen content (Fig. 1, Table 2). Again, the IgY Inhibition Assay showed a lower correlation than the CAP Inhibition Assay, although statistically significant.

Taking the modified Lowry Method as reference, the test characteristics of the three methods to measure the NRL allergen content are summarized in Table 3. According to the ROC curve analysis, the accuracy, determined by the area under the curve (AUC) of the IgY Inhibition Assay, was lower than that of the CAP Inhibition Assay and the Beezhold ELISA Inhibition Assay. At the optimal cut-off the CAP Inhibition Assay and the Beezhold ELISA Inhibition Assay were similar. The strength of agreement was very good between the CAP Inhibition Assay and the Beezhold ELISA Inhibition Assay (kappa = 0.80; 95 % CI 0.54–1.00) and good between the IgY Inhibition Assay and the other analytical methods (kappa 0.60; 95 % CI 0.26–0.94).

4 Discussion

Allergic reactions to NRL proteins are known since the 1920s, when the allergy became a well-recognized health problem particularly among HCW due to the frequent and prolonged use of protective gloves (Turjanmaa et al. 1996). In addition, powdered NRL gloves have also been a major carrier of NRL aeroallergens leading to respiratory allergies (Heilman et al. 1996). In order to prevent sensitization and clinical relevant allergies to NRL, national and international recommendations regulate the use of NRL gloves in the health care setting. In Germany, the technical standard for handling occupational hazards TRGS 401 bans the use of powdered NRL gloves and recommends a maximum protein content of 30 μg/g glove material for the health care setting (BAuA 2008). Different methods to reduce the protein content of NRL materials, like leaching and enzyme treatment have been developed (Perrella and Gaspari 2002). Indeed, the incidence of NRL allergy (latex allergic skin diseases and especially respiratory diseases) among HCW has decreased exponentially since the implementation of the recommendations (Latza et al. 2005).

Nevertheless, nowadays the general incidence of NRL allergy seems to have reached a steady state without a further decrease (Boonchai et al. 2014; Merget et al. 2010). Meanwhile, the use of NRL gloves is very common also in other industrial sectors, like the food industry, hair dressing, and cleaning, as well as in glove manufacturing (Sanguanchaiyakrit et al. 2014). The current regulations to limit allergenicity of NRL gloves protect only HCW, witnessing a lack

Table 1 Results of protein and natural rubber latex (NRL) allergen content

Glove ID	Type	Extracted weight (g)	Total protein content (Lowry) (μg/g)[a]	Latex allergen content (μg/g glove material)		
				CAP Inhibition Assay	Beezhold ELISA Inhibition Assay[b]	IgY Inhibition Assay[c]
1	PF	6.30	<LOD	15.90	<LOD	85.37
2	PF	8.60	1304.65	478.15	273.75	268.96
3	PF	12.30	<LOD	2.52	<LOD	<LOD
4	PF	6.90	<LOD	4.10	<LOD	<LOD
5	PF	7.01	<LOD	1.39	<LOD	<LOD
6	PF	7.85	<LOD	12.10	<LOD	<LOD
7	PF	14.10	1207.34	868.90	87.75	581.98
8	PF	5.40	1080.00	129.77	33.14	34.39
9	PF	6.47	<LOD	3.47	<LOD	<LOD
10	PF	5.82	<LOD	9.39	<LOD	<LOD
11	PF	6.48	<LOD	31.96	<LOD	<LOD
12	PF	6.39	<LOD	3.37	<LOD	<LOD
13	PF	5.74	974.00	67.34	32.76	11.71
14	P	6.68	<LOD	30.17	11.15	<LOD
15	P	5.27	<LOD	16.35	<LOD	<LOD
16	P	5.89	<LOD	16.60	6.35	<LOD
17	P	6.10	1057.38	88.00	40.60	<LOD
18	P	5.63	195.38	55.25	44.55	179.59
19	P	5.70	215.00	30.51	17.89	71.18
20	P	5.85	260.00	48.43	9.85	35.60

P powdered, *PF* powder free, *LOD* limit of detection
[a]LOD for modified Lowry Method = 10 μg/g
[b]LOD for Beezhold ELISA Inhibition Assay = 5 μg/g
[c]LOD for IgY Inhibition Assay = 10 μg/g

Table 2 Correlation between the different analytical methods

Correlation	r_s	p
Modified Lowry Method – CAP Inhibition Assay	0.866	<0.01
Modified Lowry Method – Beezhold ELISA Inhibition Assay	0.883	<0.01
Modified Lowry Method – IgY Inhibition Assay	0.735	<0.01
Beezhold ELISA Inhibition Assay – CAP Inhibition Assay	0.892	<0.01
Beezhold ELISA Inhibition Assay – IgY Inhibition Assay	0.674	<0.01
CAP Inhibition Assay – IgY Inhibition Assay	0.672	<0.01

of awareness to the risk of NRL allergy in other professions in which NRL gloves are also frequently used. For these industrial and service sectors there are no regulations or recommendations in place regarding the protein or allergen and powder content of gloves, the same is true for the general public. We found a relatively high amount of gloves with high protein and NRL allergen content within the 20 randomly bought latex gloves. These gloves, available to the general public in drugstores, pharmacies, etc., bear the risk of NRL sensitization and consequently of skin allergic reactions. Above all, nearly half of the gloves with high protein content were powdered, posing a risk to the development of allergic asthma.

A further problem is the measurement of the NRL allergen content in NRL materials. A number of different test procedures have been described. Due to its simplicity, the modified Lowry Method is commonly in use and it is the standard procedure in Germany (Deutsches

Fig. 1 Correlation between tests results

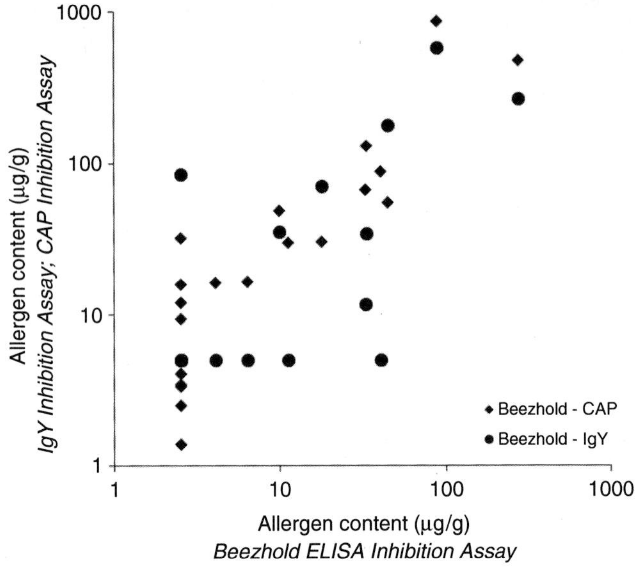

Table 3 Characteristics of tests to measure the natural rubber latex (NRL) allergen content using the modified Lowry Method as reference

	Cut-off values (µg/g)	Sensitivity (%)	Specificity (%)	AUC
CAP Inhibition Assay	10	100	50	
	30	100	83	0.99
Bezzhold ELISA Inhibition Assay	5	100	83	
	10	88	92	
	30	75	100	0.99
IgY Inhibition Assay	10	88	92	
	30	75	92	0.89

AUC area under the curve

Institut für Normung e.V. 2007). A fundamental limitation of this method is that it measures total proteins instead of NRL allergens under the assumption that both positively correlate. Indeed, several authors question this assumption and refuse the use of the modified Lowry Method as a marker for allergenicity of NRL materials (Palosuo et al. 2002). Available methods to measure the NRL allergen content are the CAP Inhibition Assay and the Beezhold ELISA Inhibition Assay, among others. The advantages of the CAP Inhibition Assay are the timesaving and little interference-prone semi-automated ImmunoCAP®-method. Unfortunately, the CAP Inhibition Assay requires the use of sera from NRL allergic subjects, which are naturally finite, not standardized, with variable antibody patterns and levels. Thus, the reproducibility of the method has been questioned, since the spectrum of allergens covered varies depending on the pool of sera (Koch 2002). It has been proposed to enhance the standardization of the CAP Inhibition Assay by measuring Hev b 1 as a marker allergen, since Hev b 1 is one of the known major allergens (Raulf-Heimsoth et al. 2000). However, the use of marker allergens for allergen content description is again questionable, since Hev b 1 is not always present in NRL gloves (Mabe et al. 2009).

In order to cover the whole spectrum of NRL allergens, Beezhold developed an inhibition assay with NRL IgG-antibodies from rabbits immunized with the complete spectrum of latex proteins (ASTM D 6499) (Tomazic-Jezic et al. 2002). The method has, however, important disadvantages. First of all, it requires to sacrifice the animals (rabbits) in order to obtain the anti-NRL-antibodies for the assay. In addition, the amount and the spectrum of antibodies produced by the rabbits is also frequently limited (Schade et al. 2001). Also, IgG antibodies are known to tend to react non-specifically, leading to false positive results (Hansson et al. 2008). Finally, this method is time-consuming and cost-intensive.

Our aim was to develop an assay overcoming the disadvantages of the above mentioned methods. We produced an assay with hen IgY-antibodies. The production of antibodies in the eggs of immunized hens was already described in 1893 (Klemperer 1893). With the growing sensibility for animal protection and increasing legal restrictions to the use of animals in research, the extraction of IgY-antibodies from hen eggs gained academic interest in the 1990s. The method has several advantages. First, immunization of the hen elicits IgY-immunoreactivity reliably. Second, unspecific reactions of IgY antibodies are known to be unlikely (Schade et al. 2001). Third, after immunization, the eggs laid by the hen will contain the appropriate antibodies lifelong. With this method, no animals need to be sacrificed, thus the method raises no major ethical concerns related to animal protection issues in contrast to the Beezhold ELISA Inhibition Assay. In addition, the antibody production is much more effective than in rabbits; the antibody content in one egg is approximately the same that the one in the whole blood of one rabbit (Wooley and Landon 1995). Finally, the antibodies can be stored simply for years by storing the refrigerated undamaged egg.

Our results demonstrate that the new IgY based test assay correlates well with the hitherto widespread methods such as the modified Lowry Method, the CAP Inhibition Assay, and the Beezhold ELISA Inhibition Assay. Indeed, in comparison to the other test methods, sensitivity and specificity of the IgY assay was high, although lower than that of the other tests. The IgY Inhibition Assay also showed a good degree of agreement with the other tests. Overall, test accuracy of our assay was good.

Considering the advantages and disadvantages of the heretofore tested procedures and in view of test accuracy, our results confirm that the cost-effective and easy to perform the modified Lowry Assay is the test of choice for preventative purposes. As described earlier (EN 455-3), we saw a good correlation of protein and allergen content in NRL gloves. Particularly, if the modified Lowry Assay delivers results under the limit of detection, it can be reliably inferred that the NRL allergen content would be at very low levels, and thus risk of sensitization would also be low. Nevertheless, in situations where there is a need to specifically measure the whole range of NRL allergens instead of total proteins, we recommend the use of the IgY Inhibition Assay. In our view, that test allows a respectful and considerate use of animals while achieving a satisfactory accuracy.

In conclusion, our results indicate that there is a need to extend the restrictions of allergen contents, ideally to all available NRL gloves. Otherwise, further decreases in the incidence of latex allergy will hardly be achieved.

Acknowledgements We would like to thank Frauke Koops for her skillful technical assistance.

Conflicts of Interest The authors declare no conflicts of interest regarding this publication.

References

BAuA Technische Regel für Gefahrstoffe 401 (TRGS 401) (2008) Available from: http://www.baua.de/de/Themen-von-A-Z/Gefahrstoffe/TRGS/TRGS-401.html. Accessed 20 Dec 2015

Baur X, Chen Z, Raulf-Heimsoth M, Degens P (1997) Protein and allergen content of various natural latex articles. Allergy 52:661–664

Boonchai W, Sirikudta W, Iamtharachai P, Kasemsarn P (2014) Latex glove-related symptoms among health care workers: a self-report questionnaire-based survey. Dermatitis 25:135–139

Cabanes N, Ilgea JM, de la Hoz B, Agustín P, Blanco C, Domínguez J, Lázaro M, Lleonart R, Méndez J, Nieto A, Rodríguez A, Rubia N, Tabar A, Beitia JM,

Dieguez MC, Martínez-Cócera C, Quirce S (2012) Latex allergy: position paper. J Investig Allergol Clin Immunol 22:313–330
Deutsches Institut für Normung e.V (2007) DIN EN 455-3. Medizinische Handschuhe zum einmaligen Gebrauch. Teil 3: Anforderung und Prüfung für die biologische Bewertung: Deutsche Fassung EN 455-3: 2006
Hansson LO, Flodin M, Nilsen T, Caldwell K, Fromell K, Sunde K, Larsson A (2008) Comparison between chicken and rabbit antibody based particle enhanced cystatin C reagents for immunoturbidimetry. J Immunoassay Immunochem 29:1–9
Heilman DK, Jones RT, Swanson MC, Yunginger JW (1996) A prospective, controlled study showing that rubber gloves are the major contributor to latex aeroallergen levels in the operating room. J Allergy Clin Immunol 98:325–330
International Union of Immunological Societies Allergen Nomenclature Subcommittee (2015) Allergen nomenclature: Hevea brasiliensis. Available from: http://www.allergen.org/search.php?allergenname%20=%20andallergensource%20=%20Hevea+brasiliensisandTaxSource%20=%20andTaxOrder%20=%20andfoodallerg%20=%20allandbioname%20=. Accessed 20 Dec 2015
Klemperer F (1893) Ueber natürliche immunität und ihre verwerthung für die immunisirungstherapie. Archiv für die Experimentelle Pathologie und Pharmakologie 31:356–382
Koch HU (2002) Protein determination in latex gloves. Source of Surgery, 9, 1. Available from: http://www.ansellhealthcare.com/america/latamer/source/april02-5.htm. Accessed 20 Dec 2015
Lamberti M, Buonanno R, Ritonnaro C, Giovane G, Crispino G, Feola A, Medici A, Sannolo N, Di Carlo A, Di Domenico M (2015) Molecular profile of sensitization in subjects with short occupational exposure to latex. Int J Occup Med Environ Health 28:841–848
Latza U, Haamann F, Baur X (2005) Effectiveness of a nationwide interdisciplinary preventive programme for latex allergy. Int Arch Occup Environ Health 78:394–402
Mabe DO, Singh TS, Bello B, Jeebhay MF, Lopata AL, Wadee A (2009) Allergenicity of latex rubber products used in South African dental schools. S Afr Med J 99:672–674
Merget R, van Kampen V, Sucker K, Heinze E, Taeger D, Goldscheid N, Haufs MG, Raulf-Heimsoth M, Kromark K, Nienhaus A, Bruening T (2010) The German experience 10 years after the latex allergy epidemic: need for further preventive measures in healthcare employees with latex allergy. Int Arch Occup Environ Health 83:895–903
Mota AN, Turrini RN (2012) Perioperative latex hypersensitivity reactions: an integrative review. Rev Lat Am Enfermagem 20:411–420
Palosuo T, Alenius H, Turjanmaa K (2002) Quantitation of latex allergens. Methods 27:52–58
Perrella FW, Gaspari AA (2002) Natural rubber latex protein reduction with an emphasis on enzyme treatment. Methods 27:77–86
Raulf-Heimsoth M, Sander I, Chen Z, Borowitzki G, Diewald K, van Kampen V, Baur X (2000) Development of a monoclonal antibody-based sandwich ELISA for detection of the latex allergen Hev b 1. Int Arch Allergy Immunol 123:236–241
Sanguanchaiyakrit N, Povey AC, de Vocht F (2014) Personal exposure to inhalable dust and the specific latex aero-allergen, Hev b6.02, in latex glove manufacturing in Thailand. Ann Occup Hyg 58:542–550
Schade R, Behn I, Erhard M, Hlinak A, Staak C (2001) Chicken egg yolk antibodies, production and application. Springer, Berlin/Heidelberg/New York
Tarlo SM, Sussman G, Contala A, Swanson MC (1994) Control of airborne latex by use of powder-free latex gloves. J Allergy Clin Immunol 93:985–989
Tomazic-Jezic VJ, Woohlhiser MR, Beezhold DH (2002) ELISA inhibition assay for the quantification of antigenic protein in natural rubber latex. J Immunoassay Immunochem 23:261–278
Turjanmaa K, Reunala T, Rasanen L (1988) Comparison of diagnostic methods in latex surgical glove contact urticaria. Contact Dermatitis 19:241–247
Turjanmaa K, Alenius H, Mäkinen-Kiljunen S, Reunala T, Palosuo T (1996) Natural rubber latex allergy. Allergy 51:593–602
Vandenplas O (1995) Occupational asthma caused by natural rubber latex. Eur Respir J 8:1957–1965
Wooley JA, Landon J (1995) Comparison of antibody production to human interleukine-6 (I-L6) by sheep and chickens. J Immunol Methods 178:253–265

Advs Exp. Medicine, Biology - Neuroscience and Respiration (2016) 24: 45–50
DOI 10.1007/5584_2016_228

Published online: 31 May 2016

Thermal Sensitivity and Dimethyl Sulfoxide (DMSO)

Kotaro Takeda, Mieczyslaw Pokorski, and Yasumasa Okada

Abstract

Dimethyl sulfoxide (DMSO) is commonly used as a solvent for hydrophobic substances, but the compound's innate bioactivity is an area of limited understanding. In this investigation we seek to determine the analgesic potential of DMSO. We addressed the issue by assessing the perception of thermal pain stimulus, using a 55 °C hotplate design, in conscious mice. The latency of withdrawal behaviors over a range of incremental accumulative intraperitoneal DMSO doses (0.5–15.5 g/kg) in the same mouse was taken as a measure of thermal endurance. The findings were that the latency, on average, amounted to 15–30 s and it differed inappreciably between the sequential DMSO conditions. Nor was it different from the pre-DMSO control conditions. Thus, DMSO did not influence the cutaneous thermal pain perception. The findings do not lend support to those literature reports that point to the plausible antinociceptive potential of DMSO as one of a plethora of its innate bioactivities. However, the findings concern the mouse's footpad nociceptors which have specific morphology and stimulus transduction pathways, which cannot exclude DMSO's antinociceptive influence on other types of pain or in other types of skin. Complex and as yet unresolved neural mechanisms of perception of cutaneous noxious heat stimulus should be further explored with alternative experimental designs.

K. Takeda
Clinical Research Center, National Hospital Organization Murayama Medical Center, 2-37-1 Gakuen, Musashimurayama, 208-0011 Tokyo, Japan

Fujita Memorial Nanakuri Institute, Fujita Health University, 423 Oodori-cho, Tsu 514-1296, Mie, Japan

M. Pokorski (✉)
Clinical Research Center, National Hospital Organization Murayama Medical Center, 2-37-1 Gakuen, Musashimurayama, 208-0011 Tokyo, Japan

Public Higher Medical Professional School in Opole, 68 Katowicka St, 45-060 Opole, Poland
e-mail: m_pokorski@hotmail.com

Y. Okada
Clinical Research Center, National Hospital Organization Murayama Medical Center, 2-37-1 Gakuen, Musashimurayama, 208-0011 Tokyo, Japan

Keywords
Analgesia • Dimethyl sulfoxide • Nociception • Noxious heat • Pain • Perception • Thermal stimulus

1 Introduction

Dimethyl sulfoxide (DMSO) is best known for its capability of dissolving hydrophobic substances in neurobiological research. The compound is widely used in both *in vitro* and *in vivo* experimental routines. DMSO is of low toxicity, but it appears not neutral biologically *per se*. A recent pursuit for DMSO's properties has unraveled multifarious bioactivities on the detrimental-beneficial continuum, with the predominance of advantageous effects. These effects include neuroprotection, anti-inflammatory and anti-oxidant role (Kelava et al. 2011), with clinically proven benefits in genitourinary ailments (Shirley et al. 1978). DMSO is also reported to have topical analgesic effects, in particular in the skin-related nociception, which is due likely to its ease of permeation through biological barriers. Potentiation of the antinociceptive effect of topical analgesic medications has been noted in case of DMSO admixture (Stanos and Galluzzi 2013; Kumar et al. 2011).

High temperature is an archetype nociceptive stimulus in experimental investigations. Thermal skin sensitivity, particularly assessed in the hotplate routine, in which the animal displays reflex pain-behaviors, such as paw licking or body jerks, is believed to involve supraspinal structures (Gregory et al. 2013). Perception of the thermal stimulus is widely used as a measure of the analgesic power of a compound. Therefore, in the current report we set out to assess the potential of DMSO to subdue thermal pain perception. We addressed the issue by examining the response to a hotplate-applied thermal stimulus in the conscious mouse.

2 Methods

The study was approved by the Ethics Committee for Animal Care and Use of the National Hospital Organization at the Murayama Medical Center in Musashimurayama, Tokyo. Experiments were carried out in seven male C57BL/6 mice (aged 7.1 ± 1.8 weeks, weighing 22.3 ± 3.6 g) that were housed at 12/12 light-dark cycle, with the light on at 7:00 a.m., and at controlled temperature of 25 °C. The current study on the influence of DMSO on perception of thermal pain expands on the previous investigation concerning the interaction of DMSO with respiratory function and it was carried out in the same set of conscious mice (Takeda et al. 2016). Sensitivity to thermal pain was considered an independent research ramification of the DMSO bioactivity and, therefore, was herein described as a separate entity.

Thermal sensitivity to acute hotplate stimulation was studied according to the method of Anthony and Annika (2007). Briefly, the unrestrained conscious mouse was placed on a water hotplate, covered with a transparent acrylic glass cylinder to prevent from leaving the platform, preheated to 55 °C (Fig. 1). The latency time from the animal placement on the hotplate to the first instance of upright standing or jerky jump on the hind limbs was taken as a measure of thermal pain endurance. The test was discontinued at the latency time measured or at 60 s in case of lack of response. Hotplate examination was carried out in the untreated mice, physiological saline-treated mice, both considered basic controls, and then was repeated 50 min after each intraperitoneal injection of

Fig. 1 Schematic representation of the hotplate experiment

incremental doses of DMSO according to the scheme outlined below:

- Untreated
- Physiological saline 1.82 mL/kg
- DMSO 0.46 mL/kg + saline 1.36 mL/kg (DMSO dose: 0.5 g/kg)
- DMSO 0.91 mL/kg + saline 0.91 mL/kg (cumulative dose: 1.5 g/kg)
- DMSO 1.82 mL/kg (cumulative dose: 3.5 g/kg)
- DMSO 3.64 mL/kg (cumulative dose: 7.5 g/kg)
- DMSO 7.28 mL/kg (cumulative dose: 15.5 g/kg)

The dose of DMSO was titrated to the range of LD_{50} dose level (Caujolle et al. 1964; Farrant 1964). DMSO has a long-term bioactivity, persisting for hours (DMSO 2007). Therefore, the most rational way to perform the experiment was to treat the incremental doses as a cumulative dose. The cumulative dose regimen had also the advantage of each mouse being its own control along the trajectory of responses, optimalizing the study outcome. DMSO administration was in each mouse preceded by physiological saline injection in a volume of 1.82 mL/kg, used as control. The mean ± SE values of response latency were compared with those before DMSO injection using Dunnett's *t*-test. The significance level was set at $p < 0.05$.

3 Results and Discussion

In all experimental conditions, except for the 15.5 g/kg DMSO, eight or nine out of the nine mice showed a paw licking, standing or jumping response to thermal stimulation. The response latency, on average, ranged between 15 and 30 s in the majority of instances, with a fairly substantial interindividual scatter of values. Overall, there were inappreciable differences in the latency between the sequential DMSO conditions and the untreated or physiological saline-treated control level. Nor were there any appreciable differences between individual DMSO conditions (Fig. 2). At the 15.5 g/kg DMSO, which borders the LD_{50} concentration (DMSO 2007), eight out of the nine mice were immobilized, so that the latency data for this condition could not be tallied and thus are not provided in the figure. The immobilization could likely be due to the central toxic effect of a near lethal dose of a compound that easily permeates through the brain-blood barrier. In fact, DMSO neurotoxic effects attributed to the barrier disruption, with brain edema or infarct, have been observed at much smaller doses (Kleindienst et al. 2006; Windrum and Morris 2003).

The finding of the current study was that DMSO over a broad spectrum of concentration did not interfere with the perception of thermal pain. DMSO did not affect pain perception either at the low end of the dose spectrum used of 0.5–1.5 g/kg, the concentration reportedly having distinct central neurotoxicity such as causing brain ischemic episodes and encephalopathy observed during transplantation of autologous stem cells cryoprotected with DMSO in hematologic cancers (Caselli et al. 2009; Windrum and Morris 2003), or at the high end of 7.5 g/kg, seldom if ever used in biological experiments, when distorted respiratory regulation is observed (Takeda et al. 2016). The detrimental bioactivities above outlined have to

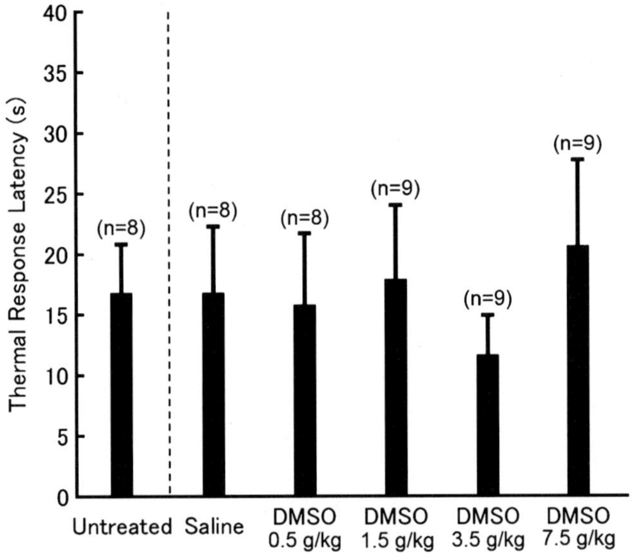

Fig. 2 Response latency to thermal stimulation of 55 °C in the untreated, physiological saline-treated, and with incremental DMSO dose-treated mice. Differences between the presented conditions were inappreciable (Dunnett's *t*-test; $p > 0.05$)

do, in all likelihood, with disruptive DMSO effect on the blood-brain barrier.

The lack of DMSO effect on thermal pain perception was a rather unexpected finding in light of its antinociceptive properties reported in some studies. DMSO has an antipain effect longer than morphine, although seemingly unrelated to opioid transmission as it is not contingent on naloxone antagonism (Haigler 1983; Haigler and Spring 1981). It also displays skin-related antinociception in topical preparations in conjuction with morphine or lidocaine (Kumar et al. 2011; Kolesnikov et al. 2000), which goes beyond that presented by either archetype analgesic agent. These antinociceptive effects might have to do with DMSO-induced dampening of peripheral C-fiber activity (Shealy 1966) or with its easy penetration into the stratum corneum when applied to the skin surface (Sulzberger et al. 1967), which makes DMSO a carrier of the accompanying analgesic compunds into deeper skin layers. DMSO applied alone, however, seems to lose its analgesic power (Kolesnikov et al. 2000).

Footpad epidermis in the mouse belongs to glabrous (non-hairy skin) which is equipped with myelinated fiber endings and unmyelinated C-fiber endings terminating freely or as Meissner's-like corpuscles among the epidermal keratinocytes (Abraira and Ginty 2013; Lindfors et al. 2006). Cutaneous nociceptors that mediate temperature sensation and pain are mostly unmyelinated endings of primary afferent neurons. Thermal sensation is transmitted *via* the dorsal root ganglions to the thalamus, where the pathway crosses to the contralateral sensory cortex (Abraira and Ginty 2013). The primary neurons are usually polymodal, responding to various sensory stimuli and containing various neuropeptides, with the predominance of calcitonin gene-related peptide and substance-P expression (Navarro et al. 1995). Thus, the neural perception pathways of thermal pain and other sensory stimuli intertwine.

The research on thermal transduction has recently focused on the transient receptor potential (TRP) channels subfamily V1 and A1. Both TRPA1 and TRPVI colocalize in neurons of the dorsal root ganglia at the lumbar level which innervate the hindpaw in the mouse (Hoffmann et al. 2013), and both are expressed in unmyelinated peripheral nerve fibers (Weller et al. 2011). Coordinated and apparently not cross-dependent action of both channel types underlies the response to suprathreshold heat stimulation, which in the mouse corresponds to >42 °C. Recent evidence gathered from TRPA1 knock-out mouse studies indicates that TRPA1 has a critical role in suprathreshold pain responsiveness specifically concerning the stimuli applied to the plantar surface of a hindpaw in the mouse (Minett et al. 2014), which is exactly what the current hotplate investigation was about. However, even in TRPA1/V1 double-knockout mice there still remains a substantial component of pain-reflex behavior (Hoffmann et al. 2013), which underscores the complexity of as yet unresolved mechanisms of noxious heat stimulation. How exactly the plantar nerve endings determine the summation of temperature intensity and duration into the above pain threshold level is by far unclear. Nonetheless, lack of thermal antinociceptive effect of DMSO suggests the compound is devoid of interaction with the specific nociceptor transducers outlined above which are responsible for transmission of the footpad thermal pain stimulus.

In conclusion, we believe we have shown that the paw-transduced behavioral withdrawal responses to noxious heat remains, on average, unchanged in DMSO-treated mice. Therefore, DMSO does not influence the cutaneous thermal pain perception. That does not exclude DMSO's antinociceptive influence on other types of pain or in other types of skin, such as hairy skin, or in other body locations; the factors that involve different neural transduction pathways and which, therefore, ought to be pursued with alternative designs of behavioral pain study.

Acknowledgements This work was supported by JSPS KAKENHI Grants 25540130, 26460311, 26670676, 15K00417, 15K12611, Health and Labor Sciences Research Grants of Japan, and JSPS Japan-Poland Researcher Exchange Program Grant.

Conflict of Interests None declared.

References

Abraira VE, Ginty DD (2013) The sensory neurons of touch. Neuron 79:618–639

Anthony WB, Annika BM (2007) Models of nociception: hot-plate, tail-flick, and formalin tests in rodents. In: Charles G, Andrew H, David S, Phil S, Susan W (eds) Current protocols in neuroscience. Wiley, New York, pp 8.9.1–8.9.16. doi:10.1002/0471142301.ns0809s41

Caselli D, Tintori V, Messeri A, Frenos S, Bambi F, Aricò M (2009) Respiratory depression and somnolence in children receiving dimethylsulfoxide and morphine during hematopoietic stem cells transplantation. Haematologica 94:152–153

Caujolle F, Caujolle D, Bouyssou H, Calvet MM (1964) Toxicity and pharmacological aptitudes of dimethylsulfoxide. C R Hebd Seances Acad Sci 258:2224–2226

DMSO (2007) Dimethyl sulfoxide. health and safety information. Bulletin #106. Gaylord Chemical Company L.L.C. Slidell, LA, USA

Farrant J (1964) Pharmacological actions and toxicity of dimethyl sulphoxide and other compounds which protect smooth muscle during freezing and thawing. J Pharm Pharmacol 16:472–483

Gregory NS, Harris AL, Robinson CR, Dougherty PM, Fuchs PN, Sluka KA (2013) An overview of animal models of pain: disease models and outcome measures. J Pain 14:1255–1269

Haigler HJ (1983) Comparison of the analgesic effects of dimethyl sulfoxide and morphine. Ann NY Acad Sci 411:19–27

Haigler HJ, Spring DD (1981) DMSO (dimethyl sulfoxide), morphine and analgesia. Life Sci 29:1545–1553

Hoffmann T, Kistner K, Miermeister F, Winkelmann R, Wittmann J, Fischer MJ, Weidner C, Reeh PW (2013) TRPA1 and TRPV1 are differentially involved in heat nociception of mice. Eur J Pain 17:1472–1482

Kelava T, Ćaver I, Čulo F (2011) Biological actions of drug solvents. Period Biol 113:311–320

Kleindienst A, Dunbar JG, Glisson R, Okuno K, Marmarou A (2006) Effect of dimethyl sulfoxide on blood-brain barrier integrity following middle cerebral artery occlusion in the rat. Acta Neurochir Suppl 96:258–262

Kolesnikov YA, Chereshnev I, Pasternak GW (2000) Analgesic synergy between topical lidocaine and topical opioids. J Pharmacol Exp Ther 295:546–551

Kumar S, Kumar S, Ganesamoni R, Mandal AK, Prasad S, Singh SK (2011) Dimethyl sulfoxide with lignocaine versus eutectic mixture of local anesthetics: prospective randomized study to compare the efficacy of cutaneous anesthesia in shock wave lithotripsy. Urol Res 39:181–183

Lindfors PH, Võikar V, Rossi J, Airaksinen MS (2006) Deficient nonpeptidergic epidermis innervation and reduced inflammatory pain in glial cell line-derived neurotrophic factor family receptor α2 knock-out mice. J Neurosci 26:1953–1960

Minett MS, Eijkelkamp N, Wood JN (2014) Significant determinants of mouse pain behaviour. PLoS One 9: e104458. doi:10.1371/journal.pone

Navarro X, Verdú E, Wendelscafer-Crabb G, Kennedy WR (1995) Innervation of cutaneous structures in the mouse hind paw: a confocal microscopy immunohistochemical study. J Neurosci Res 41:111–120

Shealy CN (1966) The physiological substrate of pain. Headache 6:101–108

Shirley SW, Stewart BH, Mirelman S (1978) Dimethyl sulfoxide in treatment of inflammatory genitourinary disorders. Urology 11:215–220

Stanos SP, Galluzzi KE (2013) Topical therapies in the management of chronic pain. Postgrad Med 125 (4 Suppl 1):25–33

Sulzberger MB, Cortese TA Jr, Fishman L, Wiley HS, Peyakovich PS (1967) Some effects of DMSO on human skin *in vivo*. Ann NY Acad Sci 141:437–450

Takeda K, Pokorski M, Sato Y, Oyamada Y, Okada Y (2016) Respiratory toxicity of dimethyl sulfoxide. Adv Exp Med Biol 885:89–96. doi:10.1007/5584_2015_187

Weller K, Reeh PW, Sauer SK (2011) TRPV1, TRPA1, and CB1 in the isolated vagus nerve – axonal chemosensitivity and control of neuropeptide release. Neuropeptides 45:391–400

Windrum P, Morris TC (2003) Severe neurotoxicity because of dimethyl sulphoxide following peripheral blood stem cell transplantation. Bone Marrow Transplant 31:315

Advs Exp. Medicine, Biology - Neuroscience and Respiration (2016) 24: 51–60
DOI 10.1007/5584_2016_226

Published online: 10 May 2016

IgE Reactivity, Work Related Allergic Symptoms, Asthma Severity, and Quality of Life in Bakers with Occupational Asthma

C. Bittner*, M.V. Garrido*, V. Harth, and A.M. Preisser

Abstract

In Germany, bakers with occupational asthma willing to stay in their job are included in an interdisciplinary program of the Social Accident Insurance for Foodstuff and Catering Industry (BGN). The primary aim is to reduce flour dust exposure, and to provide adequate medical treatment. Our aim was to evaluate the program's effect on the disease's course using routinely collected data. Forty three bakers with allergic occupational asthma and with the available baseline level of IgE (f4, f5) were investigated. Changes in IgE related to wheat and rye flour exposure were measured by ImmunoCAP test during follow-up visits. A questionnaire on work-related allergic complaints (WRAC), the Asthma Control Test (ACT), a 10-point scale of asthma severity grade, and quality of life instruments (EQ-5D-5L, Mini-AQLQ) were administered. We found an improvement of asthma severity in 88.4 % of the bakers. WRAC were reported by 65 %; 77 % had good asthma control (ACT $\geq$ 20); and 81 % had regular asthma medication. A relevant reduction of $\geq$2 CAP-classes for both allergens was seen in 12 % of the subjects. Health-related and asthma-specific quality of life was high. We conclude that satisfactory asthma control is probably the result of adequate medical management. In a subgroup of bakers with decreased specific IgE, it may also be attributed to reduced allergen exposure.

Keywords

Asthma control • Asthma related quality of life • Baker's asthma • Health related quality of life • Occupational asthma

*Both authors are contributed equally

C. Bittner (✉), M.V. Garrido*, V. Harth, and A.M. Preisser
Division of Clinical Occupational Medicine, Institute for Occupational and Maritime Medicine (ZfAM), University Medical Center Hamburg-Eppendorf, 10 Seewartenstraße, 20459 Hamburg, Germany
e-mail: cordula.bittner@bgv.hamburg.de

1 Introduction

Worldwide work-related respiratory symptoms have been known to be highly prevalent among bakery workers for decades (Baur et al. 1998), and its prevalence does not seem to be declining (Malo and Chan-Yeung 2009). Approximately 18–30 % of bakers suffer from rhinitis (Brisman et al.1995) and 5–10 % from bronchial asthma (Quirce et al. 2013). The most common cause for baker's asthma is type I allergy mediated by specific immunoglobulin E antibodies (IgE). The main allergen sources are grains, like wheat and rye, as well as baking additives (Ngahane et al. 2015; Brant 2007).

Exposure data from field studies in bakeries have shown that workers involved in the baking process as dough makers, bread formers, and bread bakers have the highest average dust and allergen exposures (Houba et al. 1997). More recent studies have shown that aside from job category, bakery size is also an important determinant of elevated exposure to flour dust allergens, with exposure increasing with bakery size (Elms et al. 2006, 2003). It has been suggested that the appropriate knowledge and the use of effective dust control measures, coupled with training and supervision, can substantially reduce flour dust exposure levels in bakeries. The overall intervention effect revealed, on average, a 50 % decrease in flour dust and wheat and rye allergen exposure. For bakers, the greatest reduction in flour dust was achieved with control measures such as the use of a mixer lid (67 %), divider oil (63 %), or focused training (54 %). However, the greatest reduction (80 %) was observed when using a combination of all control measures (Baatjies et al. 2009).

Exposure reduction measures at the work place are an alternative to resigning from bakery job for subjects with occupational asthma (Bittner et al. 2014; Preisser et al. 2011). In Germany, the Social Accident Insurance Institution for the Foodstuff and Catering Industry (Berufsgenossenschaft für Nahrungsmittel und Gastgewerbe, BGN) offers bakers, who have been diagnosed with occupational asthma but are willing to continue in their occupation, an interdisciplinary program consisting of technical and behavioural interventions for flour dust reduction and regular medical follow-up.

The aim of this study was to evaluate the effect of the program on the course of disease using routinely-collected data on IgE reactivity (as a sign of allergen exposure) and asthma control (self-assessment and physician-assessment). In addition, we studied the health-related quality of life in a subset of bakers.

2 Methods

2.1 Subjects

The study was approved by the Ethics Committee of the Hamburg Medical Association (register number PV4858). Between October 2013 and June 2015, 45 bakers (3 women and 42 men) with occupational asthma (baker's asthma) were regularly seen at our Polyclinic for Occupational Health for follow-up within the program of the BGN. 43 bakers (3 women and 40 men) with available baseline IgE (f4, f5) were included in this study. The participants gave informed consent to the analysis of routinely collected data such as specific IgE to occupational allergens, and asthma control and therapy (see below). In addition, we conducted a postal survey including health-related quality of life instruments with those who had attended our institute until October 2014 ($n = 37$).

2.2 Specific Immunoglobulin E

IgE against f4 (wheat) and f5 (rye) were measured with ImmunoCAP® (Thermo Fisher Scientific; Freiburg, Germany) at the time of the follow-up visits. For the measurement, serum samples were obtained from the participants, centrifuged at 1200 g for 15 min, aliquoted and frozen at −20 °C. The sera were analyzed at the central laboratory of the BGN in Mannheim, Germany. Results are given in CAP-Classes. Sensitization was assumed to be present if the CAP-Class was ≥ 2. We compared the CAP-class at the time of follow-up with the CAP-class reported at the time of entering the program (baseline). CAP-class changes of at least 2 classes were accepted as relevant. The baseline specific IgE was detected by ImmunoCAP® either at the central BGN laboratory or at other laboratories.

2.3 Asthma Control and Work-Related Allergic Complaints

At the follow-up visit, the bakers with occupational asthma filled out a questionnaire for the routine collection of data. They were asked to retrospectively rate the severity of their asthma complaints, both within the 4 weeks previous to the follow-up visit and before entering the intervention program, on a numerical scale, where 0 indicates "no complaints at all" and 10 the maximal severity of complaints (10-point scale of asthma severity grade). The questionnaire also included questions on the incidence of work-related allergic complaints (WRAC) such as runny nose, sneezing, nasal obstruction, and conjunctivitis, at and outside the workplace, again within a time frame of 4 weeks. In addition, the German version of the asthma-control-test (ACT) was administered. The ACT is a validated instrument consisting of five items addressing disturbance of daily activity by asthma complaints, frequency of shortness of breath, night symptoms, use of rescue medication, and a self-rated overall asthma control in the past 4 weeks (Nathan et al. 2004). The ACT score ranges from 5 to 25, where 5 represents the poorest asthma control and scores $\geq$20 are classified as a good asthma control. The minimum clinically relevant difference has been established at 3 points (Schatz et al. 2009).

Based on clinical interviews and the level of forced expired volume at 1 s (FEV_1), the attending physicians classified asthma as well controlled, partly controlled, or uncontrolled, according to the recommendations of the German National Guideline for Asthma (NVL 2009), which are modified from the Global Initiative for Asthma (GINA) assessment of asthma control. Asthma therapy was classified according to the GINA steps (step 1 = 'as-needed' of short-acting $beta_2$-agonist; SABA) (GINA 2014).

2.4 Fractional Concentration of Exhaled Nitric Oxide (FENO)

FENO levels were measured as an indicator of airway inflammation. FENO was measured with an Eco Medics™ device (Denox 88 CLD 88sp; Duernten, Switzerland), according to the recommendations of the American Thoracic Society with a flow of 50 ml/s (ATS/ERS 2005). The results are reported as ppb (parts-per-billion).

2.5 Health-Related and Asthma-Related Quality of Life (HRQoL and AQLQ)

General HRQoL was self-assessed with the EuroQoL-5D-5L instrument (Herdman et al. 2011), the German version. The instrument consists of five dimensions (mobility, self-care, usual activities, pain/discomfort, anxiety/depression) and a visual analogue scale (VAS). The participant reports his health state for each dimension on the day of completing the questionnaire, choosing among five levels: no problems, slight problems, moderate problems, severe problems, and extreme problems. The word 'problems' is replaced by 'pain or discomfort' in the fourth dimension and by 'anxious or depressed' in the fifth dimension. For each dimension, the answers are coded with a 1-digit number from '1' no problems to '5' extreme problems. The digits of the 5 dimensions are combined into a 5-digit number, which describes the overall health state of a respondent (van Reenen and Janssen 2015). For example, '11111' indicates no problems in any of the dimensions, while '21131' would indicate slight problems in mobility, moderate pain and discomfort, and no problems in the other dimensions. In the EQ VAS, the self-rated health is visualized on a 20 cm vertical, visual analogue scale with endpoints labeled 'the best health you can imagine' – 100, and 'the worst health you can imagine' – 0, and the number is marked on the scale in a box (van Reenen and Janssen 2015). The time frame for the VAS is also restricted to the day of test-taking.

Asthma-related quality of life was assessed with the German version of the Mini Asthma Quality of Life Questionnaire (MiniAQLQ), which is a validated disease-specific quality of life self-assessment instrument consisting of 15 items in four domains such as symptoms, activities, emotions, and environment over a time frame of 2 weeks. Overall score and domain scores are calculated as a mean score of the 15 items or of the respective domain items, with each score ranging from 1 to 7, where higher scores indicate better quality of life. A score change of >0.5 points is considered to be clinically relevant (Juniper et al. 1994; Juniper et al. 1999).

2.6 Statistical Analysis

Descriptive statistics are presented as median (interquartile range, IQR) or mean ± SD for continuous variables, and as frequency and percentage for categorical variables. Associations between different variables were explored with 2 × 2 contingency tables. Because of the hypothesis-generating character of the study, statistical significance was not tested.

3 Results

3.1 Study participants

Table 1 reports the characteristics of the study participants. Considerable differences were seen regarding the length of participation in the program, i.e., the years between entering the program and the actual follow-up visit, which ranged from 3.2 to 17.6 years. Twenty one percent of the participants had been less than 5 years in the program, 35 % had been 5–10 years in the program and 44 % had been 10 or more years in the program.

3.2 Self-Assessment of Asthma Control and Work-Related Allergic Complaints

In the 10-point scale of asthma severity grade, 88.4 % of the bakers reported an improvement in their asthmatic complaints of at least 1 point when retrospectively comparing their current complaints with those before entering the program. The improvements ranged from 1 point (7.0 % of the bakers) to 8 points (4.6 % of the bakers): only one baker reported a worsening of the asthma complaints in the 4 weeks previous to follow-up as compared with the time before entering the program. There was a considerable difference between the mean retrospectively self-assessed asthma control before entering the program and the 4 weeks previous to the follow-up visit. The mean improvement between both scores was 3.1 ± 2.2. Sixty five percent of the bakers reported at least one work-related allergic complaint, the most common was sneezing (51.2 %) followed by dry nasal obstruction (34.9 %).

3.3 Specific Immunoglobulin E

Table 2 shows the course of IgE CAP-classes from the baseline assessment to the follow-up visit. Thirty eight (88.3 %) of the bakers were sensitized against f4 (wheat) (CAP-class ≥2) in the baseline and 33 (76.7 %) in the follow-up. When considering f5 (rye), the level of sensitization decreased from 37 (86.0 %) in the baseline to 34 (79.0 %) in the follow-up. Between baseline and follow-up, a relevant decrease of IgE-level of two or more CAP-classes was seen in 13.9 % for f4 and in 18.6 % for f5. No change or a change of only one CAP-class for f4 and f5 was observed in 35 (81.0 %) and 33 (76.7 %), respectively. Two bakers remained non-sensitized to f4 (i.e., CAP-Class 0 or 1 at both baseline and follow-up assessments) and three remained non-sensitized to f5. Moreover, 2 of the 5 bakers who showed no sensitization at baseline, neither to f4 nor to f5, displayed

Table 1 Characteristics of study participants (n = 43)

Age at follow-up, years (mean ± SD)	47.2 ± 7.7
Female (%)	7.0
Never smokers (%)	55.8
Years in program (mean ± SD/median)	9.7 ± 4.5/9.1
Asthma control before entering the program (mean ± SD/median)	5.3 ± 2.4/6.0
Asthma control in the last 4 weeks (mean ± SD/median)	2.2 ± 2.0/2.0
ACT (% well controlled, ≥20)	76.7
Asthma control, physician assessment (% "well controlled"[a])	83.7
Allergic complaints at work (%)	
Runny nose	27.9
Sneezing	51.2
Nasal obstruction	34.9
Conjunctivitis	20.9
Asthma Therapy at follow-up (%)[b]	
As-needed SABA (step 1)	18.6
Low-dose ICS (step 2)	25.6
Low-dose ICS + LABA (step 3)	37.2
Medium/high ICS + LABA (step 4)	18.6
Medium/high ICS + LABA + oral corticosteroids/anti-IgE (step 5)	0.0
FENO at follow-up visit, ppb (mean ± SD/median)	25.2 ± 21.7/16.5
FENO at follow-up visit (% < 25 ppb)	67.0

ACT asthma control test, *FENO* fractional concentration of exhaled nitric oxide, *ICS* inhaled corticosteroids, *LABA* long-acting beta$_2$-agonists, *SABA* short-acting beta$_2$-agonists, *SD* standard deviation
[a]According to German National Asthma Guideline (NVL 2009)
[b]GINA therapy steps (GINA 2014)

detectable sensitization for both allergens (i.e., 4.7 % of the bakers).

3.4 FENO Measurements

As shown in Table 1, 67 % of the bakers had a normal FENO of less than 25 ppb. Altogether, the mean was 25.2 ppb. Increased FENO results correlated with neither the allergenic reactivity nor asthma control and quality of life.

3.5 Health-Related Quality of Life

The quality of life survey was conducted in October 2014 and was completed by 20 of the 37 (54 %) approached bakers. All but one participant answered the EQ-5D questionnaire. Approximately one third (35 %) of the participating bakers rated their health status as 'without problems' in all of the 5 dimensions (i.e., 11111). The remaining reported at least 'slight' problems in at least one dimension. The dimension for which problems were most frequently reported was 'pain/discomfort', with 40 % reporting 'slight' or 'moderate' problems in this dimension. "Moderate" problems were reported by 30 % of the bakers in the dimension 'anxiety/depression'. The mean EQ5D VAS score was 81.1 ± 11.9; the lowest score was 50, the highest 99.

3.6 Asthma-Related Quality of Life

Asthma-related QoL results are shown in Table 3. Overall, the bakers reported little asthma-related impairments in their quality of life (mean overall score 6.2 ± 0.8). The greatest impairments were reported in the dimension "environmental stimuli", i.e., impairments associated with the exposure to environmental triggers. The question related to asthma-symptom triggering by

Table 2 ImmunoCAP® results (CAP-classes) at baseline and at follow-up

f4 (wheat)			f5 (rye)			
Baseline	Follow-up	Difference	Baseline	Follow-up	Difference	Time (yr)
0	0	0	0	0	0	11.3
0	0	0	0	0	0	6.7
0	1	1	0	1	1	3.3
0	2	2	0	2	2	4.4
0	2	2	0	3	3	3.6
2	1	−1	2	0	−2	5.9
2	1	−1	2	1	−1	4.2
2	1	−1	2	1	−1	4.0
2	1	−1	1	1	0	3.3
2	1	−1	2	2	0	12.3
2	2	0	3	1	−2	17.4
2	2	0	3	2	−1	3.2
2	2	0	2	2	0	14.2
2	2	0	2	2	0	13.8
2	2	0	2	2	0	9.7
2	2	0	3	3	0	4.4
2	2	0	2	2	0	5.3
2	2	0	2	2	0	7.4
2	2	0	2	2	0	10.0
2	2	0	2	3	1	13.2
2	3	1	2	3	1	7.6
3	0	−3	3	0	−3	5.8
3	2	−1	4	2	−2	16.8
3	2	−1	3	2	−1	16.8
3	2	−1	3	3	0	7.7
3	2	−1	3	3	0	9.1
3	3	0	4	3	−1	4.3
3	3	0	3	3	0	17.6
3	3	0	3	3	0	14.0
3	3	0	3	3	0	12.4
3	3	0	3	3	0	11.8
3	3	0	4	4	0	17.1
3	3	0	3	3	0	7.5
3	3	0	3	3	0	6.8
3	4	1	4	5	1	8.6
4	1	−3	3	1	−2	12.6
4	2	−2	4	2	−2	16.5
4	2	−2	4	2	−2	12.6
4	2	−2	3	2	−1	12.9
4	4	0	5	4	−1	9.1
4	4	0	4	4	0	7.3
5	3	−2	5	3	−2	13.6
5	5	0	5	5	0	12.2

Table 3 Asthma-related quality of life in relationship to ACT, asthma therapy, and smoking status

	All	ACT		Asthma therapy		Smoking	
		<20	≥20	Step 1	Step ≥ 2	Never	Current/Former
AQLQ Dimension	n = 20	n = 4	n = 16	n = 4	n = 16	n = 10	n = 10
Symptoms	6.1 ± 0.8	5.9 ± 1.4	6.2 ± 0.6	6.2 ± 0.7	6.1 ± 0.8	6.3 ± 0.8	5.9 ± 0.8
Activity limitations	6.3 ± 0.9	5.8 ± 1.1	6.5 ± 0.8	6.4 ± 0.8	6.3 ± 0.9	6.2 ± 1.1	6.5 ± 0.5
Emotional function	6.3 ± 0.9	6.0 ± 1.1	6.4 ± 0.9	6.4 ± 1.2	6.3 ± 0.9	6.5 ± 11.0	6.1 ± 0.9
Environmental stimuli	6.0 ± 1.3	5.3 ± 2.1	6.2 ± 0.9	6.3 ± 0.7	5.9 ± 1.3	5.7 ± 1.5	6.3 ± 0.6
Overall score	6.2 ± 0.8	5.8 ± 1.3	6.3 ± 0.7	6.3 ± 0.7	6.2 ± 0.8	6.2 ± 1.0	6.2 ± 0.6

Data are means ±SD; *ACT* asthma control test, *AQLQ* asthma quality of life questionnaire

Table 4 Asthma-related quality of life in relationship to wheat and rye IgE CAP-class, and to FENO

	All	f4 CAP class course		f5 CAP class course		FENO	
		=, ⇧	⇩	=, ⇧	⇩	<25 ppb	≥25 ppb
AQLQ Dimension	n = 20	n = 15	n = 5	n = 15	n = 5	n = 14	n = 6
Symptoms	6.1 ± 0.8	6.2 ± 0.9	5.9 ± 0.4	6.2 ± 0.9	6.0 ± 0.4	6.2 ± 0.7	6.0 ± 1.0
Activity limitations	6.3 ± 0.9	6.3 ± 1.0	6.6 ± 0.3	6.2 ± 1.0	6.8 ± 0.2	6.2 ± 1.0	6.5 ± 0.5
Emotional function	6.3 ± 0.9	6.2 ± 1.0	6.5 ± 0.8	6.2 ± 1.0	6.5 ± 0.8	6.2 ± 1.0	6.5 ± 0.8
Environmental stimuli	6.0 ± 1.2	6.0 ± 1.3	6.0 ± 0.6	6.0 ± 1.3	6.0 ± 0.6	5.9 ± 1.3	6.2 ± 0.8
Overall score	6.2 ± 0.8	6.2 ± 0.9	6.2 ± 0.3	6.2 ± 0.9	6.3 ± 0.3	6.2 ± 0.9	6.3 ± 0.7

Data are means ±SD; *AQLQ* asthma quality of life questionnaire, *FENO* fractional concentration of exhaled nitric oxide, *ppb* parts per billion, = no change, ⇧ increase of ≥2 CAP-classes, ⇩ decrease of ≥2 CAP-classes

exposure to dust (Question 2 of the MiniAQLQ) was rated as the lowest, with a score of 5.2 ± 1.5. The bakers reported very few limitations in their activities. Limitations were greater for strenuous physical activities, like exercise (Question 12; 5.9 ± 1.2) and on-the-job tasks (Question 15; 6.2 ± 1.1). The asthma-related impairment of QoL was greater among those bakers who had more severe asthma (i.e., ACT < 20).

Bakers receiving treatment with "as-needed SABA" (Step 1 according to GINA) reported slightly better asthma-related quality of life than those with continuous medication. Again, the biggest differences were observed for the dimension "environmental stimuli".

In relation to CAP-class courses, there were no major differences in asthma-related quality of life between bakers with relevant decrease in the CAP-class and those without improvement (Table 4). Again, the greatest differences were observed in the domain "environmental stimuli", where bakers with no change or an increase in the CAP-class of f4 or f5 showed stronger impairments than the bakers with a relevant decrease in circulating IgE.

4 Discussion

In general, people exposed to specific occupational sensitizers, commonly high-molecular-weight agents, are at risk of developing IgE-mediated allergies, particularly rhinoconjunctivitis and bronchial asthma (Tarlo and Lemiere 2014). Baking products such as flour and baking additives, as well as vermin-infested bakeries, carry a high sensitization potential and, as a result, baker's rhinitis and baker's asthma are common occupational diseases.

Although the fundamental strategy in the primary prevention of allergic occupational asthma is the elimination of exposure to sensitizers, e.g., replacement with non-sensitizers, this is not feasible in all industries. That is the case for bakeries, where the use of sensitizing agents remains unavoidable. Exposure reduction achieved with technical and behavioral interventions can reduce the proportion of newly-sensitized workers, thus contributing to the primary prevention of occupational allergic diseases (Heederik et al. 2012).

After the onset of occupational asthma, the intervention of choice to modify the course of the disease is the complete avoidance of exposure (Tarlo et al. 2008; Nicholson et al. 2005). However, complete elimination of the hazard from the workplace or relocation to an allergen-free work area within the same enterprise is often not feasible for bakers – particularly in small family businesses, as is frequently the case in Germany – and cessation of exposure is often only achieved with resignation from the job. Studies on the socio-economic impact of occupational asthma have shown that complete avoidance can lead to severe financial consequences, either from job loss or employer change (Vandenplas and D'Alpaos 2010). When elimination of exposure is unfeasible, or jobs without exposition are unavailable, clinical practice guidelines have acknowledged that reduction of exposure may be considered an alternative in order to minimize the socio-economic impact of occupational asthma (Nicholson et al. 2005). This option is however controversial. Few adequate studies have addressed the course of disease following exposure reduction (e.g., of isocyanates and latex) compared with its complete elimination. The results of a meta-analysis have shown a higher risk of worsening asthma symptoms (OR 10.23, 95 % CI 2.97–35.28), a lower likelihood of improvement (OR 0.16, 95 % CI 0.03–0.91) and of recovery from asthma symptoms (OR 0.30, 95 % CI 0.11–0.84) when compared with complete avoidance of exposure. Furthermore, a higher risk of non-specific bronchial hyperresponsiveness has been observed (OR 5.65, 95 % CI 1.11–28.82) (Vandenplas et al. 2011). To our knowledge, no such studies have been conducted among bakers.

In Germany, since 1992 the BGN has offered an interdisciplinary program for bakers choosing to remain at their job after the onset of occupational asthma, with the aim of reducing the level of exposure (LOE) to the causative agents through technical and behavioral interventions (Grieshaber et al. 1998). To date, the effectiveness of the program in reducing the LOE has not been systematically evaluated. A comparison of field measurements of LOE at baseline and after implementing the intervention has not yet been conducted or at least has not been published.

Our work is based on accessible, routinely collected data. As there are no data of dust measurements available, we attempted to infer the changes of the LOE to wheat and rye allergens from the level of circulating IgE, based on the hypothesis that effective exposure reduction measures would lead to a decrease in the level of circulating specific serum IgE antibodies. Assuming a decrease of two or more CAP-classes as an indicator of relevant exposure reduction, our results may be interpreted as the limited effectiveness of LOE reduction. Such a CAP-class decrease was observed only in 5 (11.6 %) bakers for both allergens. In contrast, two of the five (40 %) bakers which had no detectable serum IgE levels to either f4 or f5 at baseline developed sensitization (i.e., CAP ≥ 2) within 3–4 years in the program, suggesting a failure of the technical and behavioral interventions for dust reduction. Both had a low ACT (11 and 20) and required intensified asthma therapy (steps 4 and 3, respectively). In the group with constant or worsening serum IgE levels, the proportion of bakers with further allergic symptoms such as rhinitis and/or conjunctivitis was slightly higher (71.8 % and 42.8 %, respectively). Accordingly, asthma control and the QoL in this group were slightly lower. These observations may validate the assumption that changes in IgE levels reflect the effectiveness of exposure reduction interventions. Indeed, monitoring of the LOE has not yet been performed. Further investigations with personal air measurements of flour dust are needed in order to test this hypothesis.

Nevertheless, on the 10-point asthma severity scale only one baker reported a worsening of condition. Four bakers did not report any change and the rest reported improvements, independent of their circulating IgE levels. However, the fact that the asthma severity grade for the time period before entering the program was assessed retrospectively, i.e., at the follow-up visit, must be taken into account, as recall bias may have led to an overestimation of the difference in asthma severity between the recalled baseline and the time-of-study-current severity. When considering the fact that rate of good asthma control at follow-up was high – 76.7 % reached an ACT ≥ 20 – it seems reasonable that the severity of asthma lessened after entering the program. The adequate medical management is surely a relevant factor for good asthma control. Approximately 80 % were in treatment with at least an inhalative corticosteroid, and 55 % had an additional long-acting beta-agonist.

Overall, both HRQoL (EQ-5D) and asthma-related quality of life (AQLQ) were high, as expected in sight of the efficient asthma control. Concerning the VAS score (81.1 ± 11.9 out of a maximum of 100), HRQoL was comparable to that of German employees (82.3 (SD not reported)) (König et al. 2005). However, fewer bakers reported the best health status (i.e., 11111) when compared to the reference general German population (35.0 % *vs.* 47.5 %) (Hinz et al. 2014). In addition, asthma-related quality of life was high; the greatest impairments were observed in the dimension "environment", particularly with respect to the question related to asthma symptom-triggering by exposure to dust, again suggesting a limited control of exposure. Despite this, the results suggest very little impairment of the professional activity due to asthma. The quality of life survey may not be representative, however, as data from less than half of the group were available. Thus, our results may either underestimate or overestimate QoL. Further limitations of our study include the lack of a comparison group, the lack of ACT assessment at baseline, and the subjective retrospective assessment of asthma severity for the time period before entering the program. These are consequences of the fact that only routinely collected data from an ongoing program were available. Further research is needed regarding the course of the disease (asthma control, quality of life) among bakers choosing to remain at work with exposure reduction compared to bakers who choose complete avoidance and resign from the occupation.

5 Conclusions

Our study shows that bakers with occupational asthma who decide to continue their jobs can achieve an acceptable degree of asthma control, provided they undergo regular pulmonary follow-up and adhere to treatment guidelines. In particular, it has to be considered that the wish to stay in the profession is often driven by economic constraints or lack of job alternatives. Nevertheless, asthma control is not convincingly associated with a regulation of allergic reactivity and allergic symptoms, like rhinitis and conjunctivitis, are not fully controlled under further exposure. When considering the monitored course of circulating IgE levels and the continued presence of allergic symptoms, our results indirectly suggest that sufficient allergen reduction in the workplace is not being achieved in a considerable proportion of the bakers. We therefore conclude that, for bakers who are unwilling to give up their current employment, there is a need to monitor the level of exposure with field air measurements. This would allow to better assess whether the technical and behavioral interventions are successful in reducing exposure and, depending on the results, to react accordingly.

Acknowledgements We would like to thank the Clinical Occupational Medicine Team of the institute for implementation and backing the patient examinations. Also we thank the BGN for the disclosure of the data, particularly Dr. Claus Hölzel and Bettina Simonis. We thank Rosalie McDonough for reviewing the manuscript.

Conflicts of Interest CB, MVG, VH and AMP have been paid by the BGN according to the German medical fee schedule for conducting follow-up visits of the bakers. AMP advises the BGN in medical questions about individual cases.

References

American Thoracic Society/European Respiratory Society (2005) ATS/ERS recommendations for standardized procedures for the online and offline measurement of exhaled lower respiratory nitric oxide and nasal nitric oxide. Am J Respir Crit Care Med 171:912–930

Baatjies R, Lopata AL, Sander I, Raulf-Heimsoth M, Bateman ED, Meijster T, Heederik D, Robins TG, Jeebhay MF (2009) Determinants of asthma phenotypes in supermarket bakery workers. Eur Respir J 3:825–833

Baur X, Degens PO, Sander I (1998) Baker's asthma: still among the most frequent occupational respiratory disorders. J Allergy Clin Immunol 102:984–997

Bittner C, Harth V, Preisser AM (2014) Die Berufskrankheit 4301 und das Bäckerasthma. Atemwegs- und Lungenkrankheiten 40:138–144

Brant A (2007) Baker's asthma. Curr Opin Allergy Clin Immunol 7:152–155

Brisman SJ, Järvholm BG (1995) Occurrence of self-reported asthma among Swedish bakers. Scand J Work Environ Health 21:487–493

Elms J, Beckett P, Griffin P, Evans P, Sams C, Roff M, Curran AD (2003) Job categories and their effect on exposure to fungal alpha-amylase and inhalable dust in the U.K. baking industry. Am Ind Hyg Assoc J 64:467–474

Elms J, Robinson E, Mason H, Iqbal S, Garrod A, Evans GS (2006) Enzyme exposure in the British baking industry. Ann Occup Hyg 50:379–384

GINA (2014) From the Global strategy for asthma management and prevention, Global Initiative for Asthma. Available from: http://www.ginasthma.org/. Accessed 14 Dec 2015

Grieshaber R, Nolting HD, Rosenau C, Stauder J, Vonier J (1998) Effectiveness of secondary preventive programs for patients with asthma and rhinitis in the baking industry. Pneumologie 52:656–665

Heederik D, Henneberger PK, Redlich CA (2012) Primary prevention: exposure reduction, skin exposure and respiratory protection. Eur Respir Rev 21:112–124

Herdman M, Gudex C, Lloyd A, Janssen MF, Kind P, Parkin D, Bonsel G, Badia X (2011) Development and preliminary testing of the new five-level version of EQ-5D (EQ-5D-5L). Qual Life Res 20:1727–1736

Hinz A, Kohlmann T, Stöbek-Richter Y, Zenger M, Brähler E (2014) The quality of life questionnaire EQ-5D-5L: psychometric properties and normative values for the general German population. Qual Life Res 23:443–447

Houba R, Heederik D, Kromhout K (1997) Grouping strategies for exposure to inhalable dust, wheat allergens and α-amylase allergens in bakeries. Ann Occup Hyg 41:287–296

Juniper EF, Guyatt GH, Willan A, Griffith LE (1994) Determining a minimal important change in a disease-specific quality of life instrument. J Clin Epidemiol 47:81–87

Juniper EF, O'Byrne PM, Guyatt GH, Ferrie PJ, King DR (1999) Development and validation of a questionnaire to measure asthma control. Eur Respir J 14:902–907

König HH, Bernert S, Angermeyer MC (2005) Health status of the German population: results of a representative survey using the EuroQol questionnaire. Gesundheitswesen 67:173–182

Malo JL, Chan-Yeung M (2009) Agents causing occupational asthma. J Allergy Clin Immunol 123:545–550

Nathan RA, Sorkness CA, Kosinski M, Schatz M, Li JT, Marcus P, Murray JJ, Pendergraft TB (2004) Development of the asthma control test: a survey for assessing asthma control. J Allergy Clin Immunol 113:59–65

Ngahane BHM, Nde F, Ngomo E, Ze EA (2015) Sensitization to workplace respiratory allergens among bakery workers in Douala, Cameroon: a cross-sectional study. Allergy Asthma Clin Immunol 11:13. doi:10.1186/s13223-015-0080-2

Nicholson PJ, Cullinan P, Taylor AJ, Burge PS, Boyle C (2005) Evidence based guidelines for the prevention, identification, and management of occupational asthma. Occup Environ Med 62:290–299

NVL Nationale Versorgungsleitlinie Asthma (2009) Available at http://www.awmf.org/. Accessed 14 Dec 2015

Preisser AM, Wilken D, Baur X (2011) Das Asthma des Bäckers – Diagnostik, Therapie, Prävention. Dtsch Med Wochenschr 136:637–641

Quirce S, Diaz-Perales A (2013) Diagnosis and management of grain-induced asthma. Allergy Asthma Immunol Res 5:348–356

Schatz M, Kosinski M, Yarlas AS, Hanlon J, Watson ME, Jhingran P (2009) The minimally important difference of the Asthma Control Test. J Allergy Clin Immunol 124:719–723

Tarlo SM, Lemiere C (2014) Occupational asthma. N Engl J Med 370:640–646

Tarlo SM, Balmes J, Balkissoon R, Beach J, Beckett W, Bernstein D, Blanc PD, Brooks SM, Cowl CT, Daroowalla F, Harber P, Lemiere C, Liss GM, Pacheco KA, Redlich CA, Rowe B, Heitzer J (2008) Diagnosis and management of work-related asthma: American College of Chest Physicians consensus statement. Chest 134:1S–41S

van Reenen M, Janssen B (2015) EQ-5D-5L User guide. Basic information on how to use the EQ-5D-5L instrument. Version 2.1. Available at http://www.euroqol.org/home.html. Accessed 14 Dec 2015

Vandenplas O, D'Alpaos V (2010) Social consequences and quality of life in work related asthma. In: Sigsaard T, Heederik D (eds) Occupational Asthma. Birkhäuser, Basel, pp 271–279

Vandenplas O, Dressel H, Wilken D, Jamart J, Heederik D, Maestrelli P, Sigsgaard T, Henneberger P, Baur X (2011) Management of occupational asthma: cessation or reduction of exposure? A systematic review of available evidence. Eur Respir J 38:804–811

Advs Exp. Medicine, Biology - Neuroscience and Respiration (2016) 24: 61–70
DOI 10.1007/5584_2016_237

Published online: 30 April 2016

Effects of Selective Inhibition of PDE4 by YM976 on Airway Reactivity and Cough in Ovalbumin-Sensitized Guinea Pigs

J. Mokrý, A. Urbanová, I. Medvedová, M. Kertys, P. Mikolka, P. Kosutová, and D. Mokrá

Abstract

Phosphodiesterases (PDEs) are enzymes involved in the degradation of cAMP and cGMP. Selective PDE4 inhibitors (e.g., roflumilast) are effective in therapy of chronic obstructive pulmonary disease associated with neutrophil inflammation. The aim of this study was to evaluate the effects of a selective PDE4 inhibitor, YM976, on citric acid-induced cough, *in vivo* and *in vitro* airway smooth muscle reactivity to histamine, and on inflammatory mediators in ovalbumin-sensitized guinea pigs, with experimentally induced eosinophil inflammation. The YM976 was administered intraperitoneally at a dose of 1.0 mg/kg once daily for 7 days. Sensitization with ovalbumin led to a significant increase in the number of coughs, and *in vivo* and *in vitro* airway reactivity. Also, increased plasma levels of IL-4, IL-5, and PAF were observed, with a significant increase in the differential count of eosinophils in both blood and bronchoalveolar lavage fluid. The YM976 suppressed the number of coughs, the airway reactivity in tracheal tissue strips, and the IL-4 level. The findings indicate that PDE4 inhibition by YM976 exerts antitussive and anti-inflammatory effects in guinea pigs with ovalbumin-induced eosinophilic inflammation.

Keywords

Airway reactivity • Animal model • Citric acid • Histamine • Inflammation • Interleukins • Organ bath • Phosphodiesterase inhibitors

J. Mokrý (✉), A. Urbanová, I. Medvedová, and M. Kertys
Biomedical Center (BioMed) Martin, Division of Respirology and Department of Pharmacology, Jessenius School of Medicine in Martin, Comenius University in Bratislava, Mala Hora 4C, SK-03601 Martin, Slovakia
e-mail: mokry@jfmed.uniba.sk

P. Mikolka, P. Kosutová, and D. Mokrá
Biomedical Center (BioMed) Martin, Division of Respirology and Department of Physiology, Jessenius School of Medicine in Martin, Comenius University in Bratislava, Mala Hora 4C, SK-03601 Martin, Slovakia

1 Introduction

Bronchial asthma and chronic obstructive pulmonary disease (COPD) are characterized by chronic inflammation, cough, and bronchoconstriction. Methylxanthines such as theophylline, aminophylline, or doxophylline are used to treat these diseases. Inhibition of phosphodiesterase (PDE) has been historically considered as the major mechanism of the drugs' action. However, in the therapeutically relevant plasma concentrations several other mechanisms are involved in their effects (Mokrý and Mokra 2013), including antagonism of adenosine receptors and activation of histone-deacetylase (HDAC-2). These mechanisms of action lead to such pharmacological benefits as bronchodilation, suppression of inflammation, and an antitussive effect (Mokrý and Nosalova 2011).

Theophylline having a relatively mixed and low specific mechanism of action, leads to numerous interactions with other drugs and has a narrow therapeutic range. Administration of theophylline is often associated with the occurrence of adverse effects, limiting its use as an antitussive drug (Nosalová and Mokrý 2001). The mechanism of action of theophylline, most extensively described, is a non-selective inhibition of PDE. The enzyme is classified into 11 superfamilies of metalophosphohydrolazes, hydrolyzing cAMP and cGMP to their inactive metabolites (Bender and Beavo 2006). Some of PDE isoenzymes play an important role in the regulation of airway diameter and smooth muscle function. The expression of PDE3 and PDE4, both hydrolyzing cAMP, has been identified in airway smooth muscles and inflammatory cells of the respiratory system (Medvedová et al. 2015; Mokrý and Mokra 2013). There are other PDE isoforms such as PDE1, PDE5, or PDE7 which are involved in the regulation of cAMP and cGMP and thus in inflammation (Christensen et al. 2012). Therefore, selective inhibitors of PDEs, especially PDE3 and PDE4, or dual PDE3/4 inhibitors, have attracted increasing attention in respiratory therapy (Abbott-Banner and Page 2014; Page and Spina 2012).

The first generation of PDE4 inhibitors is represented by rolipram, which has not been introduced into clinical practice due to its adverse effects (nausea and vomiting) at higher doses required for sufficient bronchodilation and anti-inflammatory action. A second generation of PDE4 inhibitors, represented by roflumilast and cilomilast, have a better perspective, as they maintain the anti-inflammatory and immunomodulation effects with a lower incidence of adverse effects (Karish and Gagnon 2006). Roflumilast, based on successful results of several clinical studies, has been recently approved for clinical use in adult patients with severe COPD (Fabbri et al. 2010; Giembycz and Field 2010) and is currently tested for bronchial asthma (Medvedová et al. 2015). Another prospective PDE4 inhibitor, YM976 (4-(3-chlorophenyl)-1,7-diethypyrido[2,3-*d*]pyrimidin-2(1*H*)-one), appears to have bronchodilating, anti-asthmatic, and anti-inflammatory effects and little emetogenic activity, which was confirmed in several models of allergic inflammation (Moriuchi et al. 2003; Aoki et al. 2000a, 2000b; Aoki et al. 2001a, 2001b). Furthermore, YM976 has been successfully tested in a model of experimental bladder inflammation (Kitta et al. 2008).

Yet there are limited data on the antitussive effect of selective PDE4 inhibitors (Mokrý et al. 2008, 2013; Morjaria et al. 2013; Mokrý and Nosalova 2011; Fujimura and Liu 2007). Thus, the aim of this study was to elucidate the involvement of PDE4 in cough and bronchoconstriction, using the selective PDE4 inhibitor YM976 in the ovalbumin-sensitized guinea pigs.

2 Methods

The study protocol was approved by a local Ethics Committee of Jessenius Faculty of Medicine, Comenius University in Martin, Slovakia. Twenty four healthy adult male guinea pigs (250–350 g) were used for the study. They were kept in an animal house and had food and water *ad libitum*. Airway hyperresponsiveness was induced with the ovalbumin antigen in two groups of animals. Another group was used as *naïve* controls without sensitization (n = 8 each group). One of the sensitized groups was left without treatment, treated only with water for injection at a dose of 3.0 mL/kg. The animals of the other sensitized group were treated with the PDE4 inhibitor YM976 (Sigma Aldrich Chemie GmbH; Hamburg, Germany) 1.0 mg/kg, i.p., dissolved in water for injection (3.0 mL/kg) for 7 consecutive days.

2.1 Antigen-Induced Airway Hyperresponsiveness

Sensitization of guinea pigs by the allergen ovalbumin was performed over a period of 14 days according to the method described previously (Franová et al. 2007; Mokrý et al. 2009). Briefly, 0.5 mL of 1.0 % ovalbumin was initially administered doubly, intraperitoneally and subcutaneously, then 1.0 mL intraperitoneally only on the 3rd day, and finally another 1.0 mL by inhalation (30 s challenge) on the 14th day of sensitization. The *in vivo* airway reactivity to mediators of bronchospasm and cough reflex assessments followed 5 h after the inhalative ovalbumin challenge, followed by *in vitro* assessments after sacrificing the animal. The drugs tested were administered 30 min before the *in vivo* assessment. The animals that demonstrated a minimum of 20 % increase in specific airway resistance after the ovalbumin challenge on the last day of sensitization were included in the further study protocol.

2.2 Cough Reflex Assessment

Chemically-induced cough was used for the assessment of cough reflex (Mokrý and Nosalova 2011; Sutovska et al. 2009). For cough provocation citric acid aerosol at a concentration of 0.6 mol/L in saline was introduced into the nasal chamber of a double chamber plethysmograph (Hugo Sachs; March, Germany). The inhalation of citric acid lasted for 2 min. During this time and the following 2 min, two well-trained observers evaluated visually the number of cough efforts. To distinguish cough from sneezing or movement artifacts, subsequent evaluation of the computer records of airflow in the nasal chamber was performed and verified against the video records.

2.3 Evaluation of *in vivo* Airway Reactivity

Specific airway resistance was considered an indicator of the *in vivo* airway reactivity. Changes in airway resistance were evaluated in a double chamber whole body plethysmograph (Hugo Sachs; March, Germany) in response to a short-term inhalation (2 min) of the bronchoconstrictor histamine – 10^{-6} mol/L in saline according to the method of Mokrý et al. (2008). Airway reactivity in response to saline nebulization was used as reference. The interval between histamine and saline exposures, separated by room air insufflation, amounted to 5 min.

2.4 Evaluation of *in vitro* Airway Reactivity

After sacrificing the animals, trachea and lungs were immediately excised. Tracheal strips (approximately 15 mm) were cut on the opposite side of a smooth muscle. Lung tissue strips (2 × 2 × 15 mm) were cut from the margin of upper and lower lobe of left lungs. The strips were mounted between two hooks and placed into the 20 mL organ chambers containing the Krebs-Henseleit buffer (NaCl 110.0 mmol/L, KCl 4.8 mmol/L, $CaCl_2$ 2.4 mmol/L, $MgSO_4$ 1.2 mmol/L, KH_2PO_4 1.2 mmol/L, $NaHCO_3$ 25.0 mmol/L, and glucose 10.0 mmol/L in glass-distilled water). The chambers were maintained at 36.5 ± 0.5 °C and aerated continuously with a mixture of 95 % O_2 and 5 % CO_2 to maintain pH of 7.5 ± 0.1. One of the hooks was connected to a force transducer and an amplifier, and tension changes were recorded online using computer software (Experimetria; Budapest, Hungary). Tissue strips were initially set to 4 g of tension for 30 min (loading phase). Then, tension was readjusted to a baseline value of 2 g for another 30 min (adaptation phase). During both phases, tissue strips were washed with fresh buffer at 10 min intervals. Cumulative doses of histamine or acetylcholine (10^{-8}–10^{-3} mol/L) Sigma-Aldrich Chemie GmbH; Hamburg, Germany) were added after the adaptation phase had been finished and a continuous recording of contractions was made (Mokrý et al. 2009). Data of the tracheal and lung tissue reactivity are showed in grams (g) of the smooth muscle tension.

2.5 Cell Content in Blood and Bronchoalveolar Lavage Fluid

Samples of blood were taken immediately after sacrificing the animals. Bronchoalveolar lavage (BAL) of the right lung was performed twice with pre-heated saline (37 °C) in the amount of 0.01 mL/g of body weight. A total white blood cells (WBC) count in blood was determined in Bürker's chamber after staining with Türck's solution. Differential leukocyte count in blood and BAL fluid was evaluated microscopically after panchromatic May-Grünwald/Giemsa-Romanowski staining and relative counts of neutrophils, eosinophils, and lymphocytes in blood and monocytes, neutrophils, and eosinophils in BAL fluid were determined (in %).

2.6 Biochemical Assay

Biochemical markers in the blood plasma were measured by standard ELISA method with commercially available kits for guinea pigs: interleukin (IL)-4 (Cusabio Life Sciences; Wuhan, China), IL-5 (USCN Life Sciences Inc.; Wuhan, China), and platelet activating factor (PAF) (BlueGene Biotech Co; Shanghai, China), all expressed in pg/mL.

2.7 Statistical Analysis

Data are shown as means ± SE. Statistical differences were assessed with one-way ANOVA with post-hoc LSD test was used. A p-value of <0.05 was used to define significant differences.

3 Results

Guinea pigs sensitization with ovalbumin led to a significant increase in the number of cough efforts. Both single dose and repeated (7 days) doses of the PDE4 inhibitor YM976 at a dose of 1.0 mg/kg i.p. caused a significant drop in the number of coughs compared with ovalbumin-sensitized guinea pigs treated with saline (Fig. 1). A single and repetitive dose of YM976 also caused a significant drop in specific airway resistance after histamine nebulization *in vivo* (Fig. 2). Likewise, tracheal smooth muscle contractility in response to cumulative doses of acetylcholine and histamine *in vitro* demonstrated a significant drop in the YM976 treated group (Fig. 3a and b). In contrast, changes in *in vitro* airway reactivity to cumulative doses of acetylcholine and histamine in lungs tissue strips were insignificant after 7 days of YM976 administration (Fig. 4a and b).

A significant increase in the eosinophils count observed in the blood and BAL fluid after sensitization in the untreated group confirmed the appearance of eosinophilic inflammation due to the sensitization protocol employed in the study. Administration of YM976 for 7 consecutive days appreciably suppressed the percentage of eosinophils in both blood and BAL fluid. These changes were associated with an increase in the number of monocytes in the BAL fluid (Table 1). Plasma levels of IL-4, IL-5, and PAF appreciably increased in the sensitized untreated group of guinea pigs. However, YM976 administered for 7 days suppressed only the level of IL-4 (Table 2).

4 Discussion

PDEs are known to influence airway inflammation and airway smooth muscle contractility. Selective inhibitors of PDE have been extensively studied for their anti-inflammatory action.

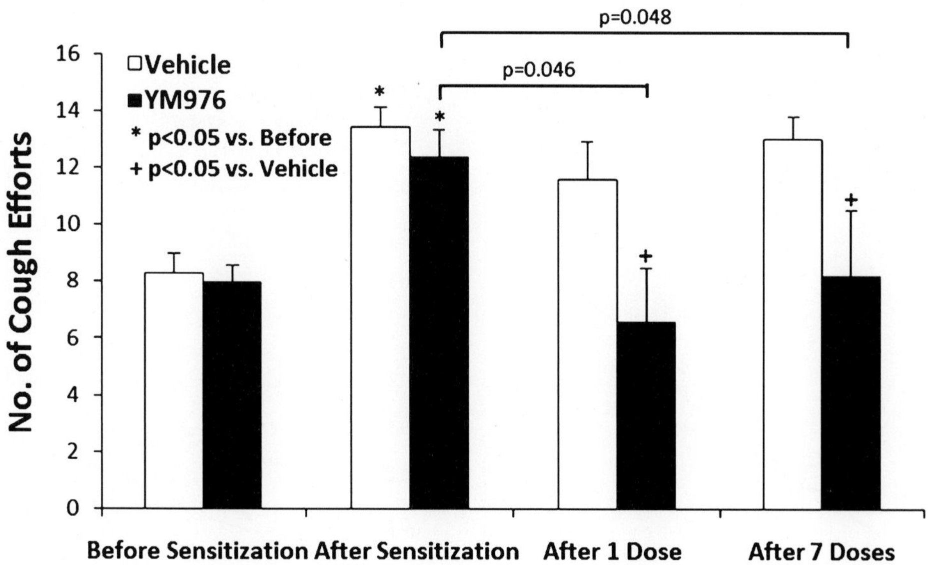

Fig. 1 Number of cough efforts registered for 2 min during and 2 min after inhalation of citric acid aerosol in healthy and ovalbumin-sensitized guinea pigs before and after pre-treatment with a single or repetitive dose (for 7 days) of saline vehicle and the PDE4 inhibitor YM976 (1.0 mg/kg, i.p.)

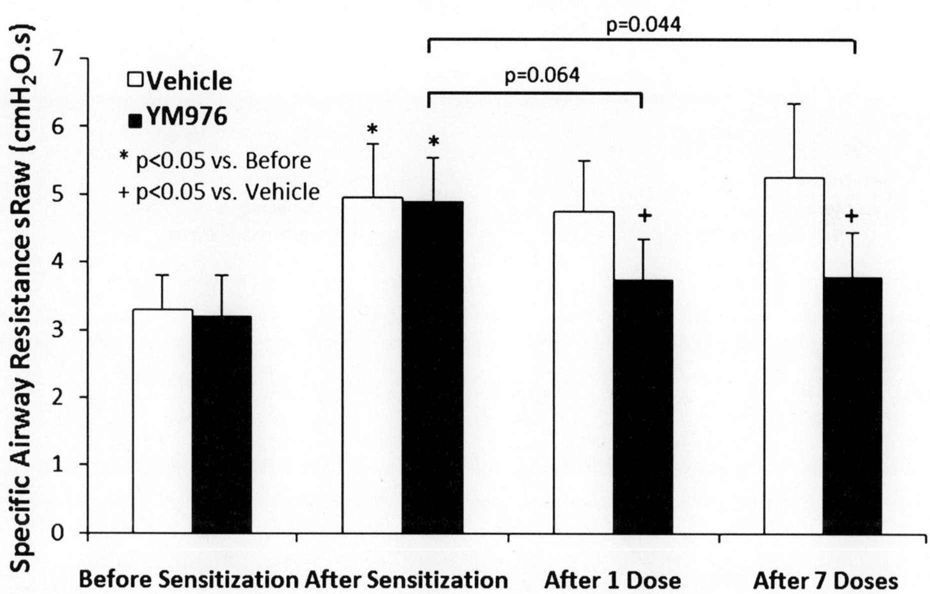

Fig. 2 Specific airway resistance measured after inhalation of histamine (10^{-6} mol/L) in healthy and ovalbumin-sensitized guinea pigs before and after pre-treatment with a single or repetitive (for 7 days) dose of saline vehicle and the PDE4 inhibitor YM976 (1.0 mg/kg, i.p.)

The PDE3, which is expressed in airway smooth muscles, apart from myocardium, vessels, and gastrointestinal tract, seems to clearly influence airway hyperresponsiveness and cough (Fujimura and Liu 2007). However, inhibitors of PDE4 are considered even more powerful in suppression of inflammation. Thus, therapeutic use of selective (PDE3, PDE4, PDE7) or dual (PDE3/4, PDE4/7) PDE inhibitors could have a beneficial effect in disorders that manifest cough

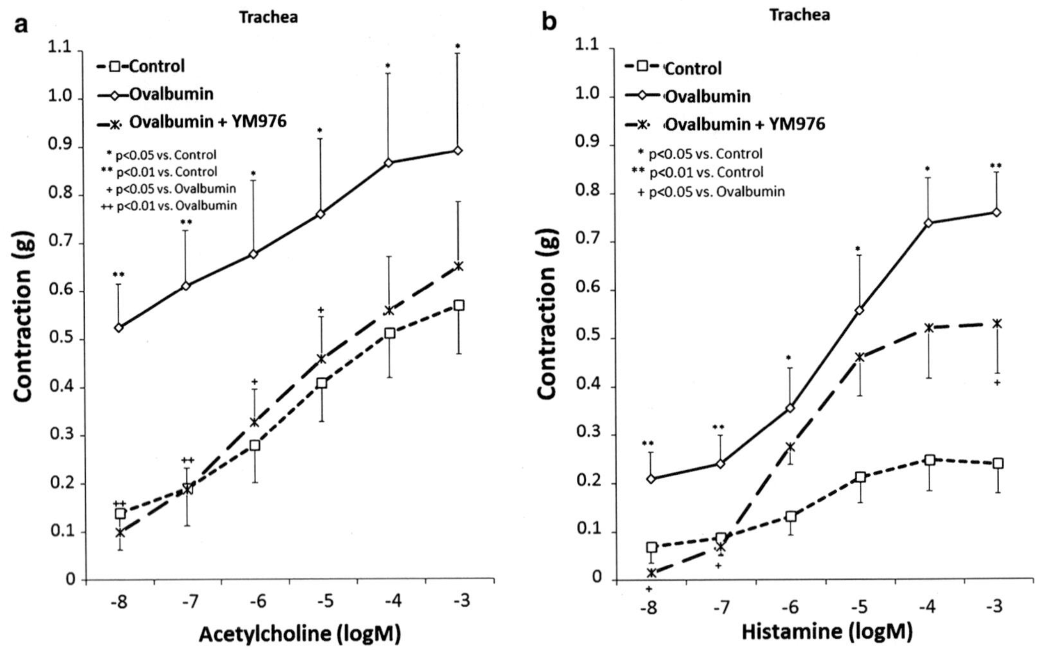

Fig. 3 *In vitro* airway reactivity of tracheal tissue strips to acetylcholine (**a**) and histamine (**b**) in healthy and ovalbumin-sensitized guinea pigs after 7-day long pre-treatment with saline vehicle and the PDE4 inhibitor YM976 (1.0 mg/kg, i.p.)

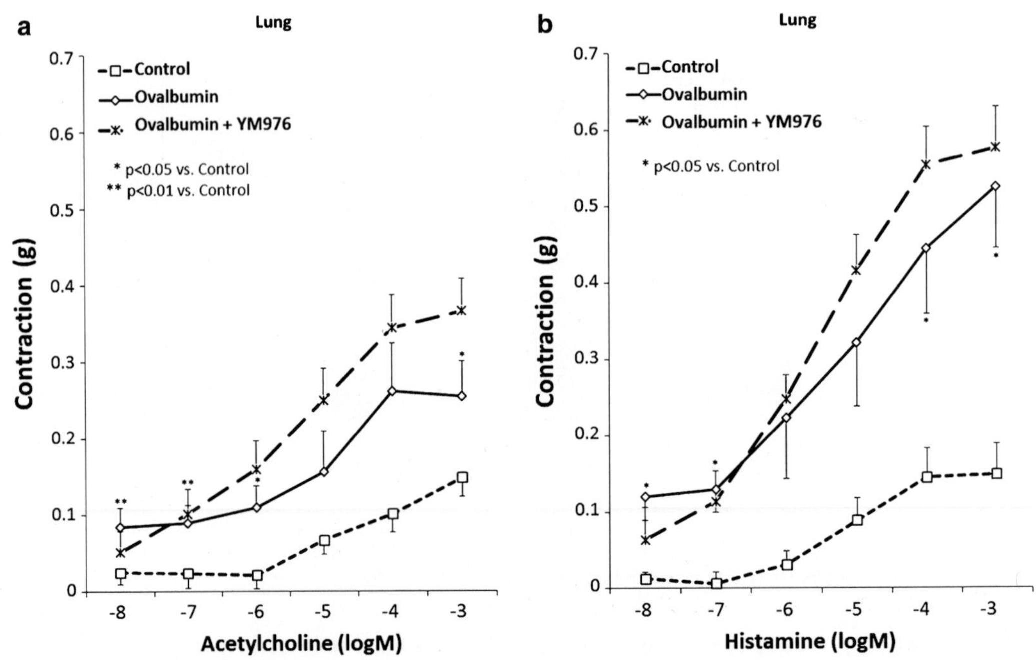

Fig. 4 *In vitro* airway reactivity of lung tissue strips to acetylcholine (**a**) and histamine (**b**) in healthy and ovalbumin-sensitized guinea pigs after 7-day long pre-treatment with saline vehicle and the PDE4 inhibitor YM976 (1.0 mg/kg, i.p.)

Table 1 Absolute and differential counts of white blood cells in blood and BAL fluid from healthy (control), ovalbumin-sensitized (OVA), and OVA + YM976-treated guinea pigs

Group	WBC blood ($\times 10^9$)	Lymph blood (%)	Neut blood (%)	Eosin blood (%)	Mono BAL (%)	Neut BAL (%)	Eosin BAL (%)
Control	2.1 ± 0.1	64.2 ± 1.8	33.9 ± 1.8	0.7 ± 0.3	95.4 ± 0.8	3.4 ± 0.4	1.2 ± 0.4
OVA	$3.4 \pm 0.5^{**}$	59.8 ± 3.5	36.9 ± 3.6	$2.2 \pm 0.3^{**}$	$60.6 \pm 2.6^{***}$	6.7 ± 1.9	$32.6 \pm 3.3^{***}$
OVA + YM976	3.9 ± 0.5	51.2 ± 8.8	46.1 ± 9.3	$0.8 \pm 0.2^{++}$	$78.5 \pm 5.4^{++}$	7.7 ± 2.2	$13.8 \pm 4.5^{++}$

Data are means ± SE

WBC absolute number of white blood cells, *Lymph* lymphocytes, *Neut* neutrophils, *Eosin* eosinophils, *Mono* monocytes, *BAL* bronchoalveolar lavage fluid

$^{**}p < 0.01$ vs. control, $^{***}p < 0.001$ vs. control, $^{++}p < 0.01$ vs. ovalbumin

Table 2 Plasma levels of PAF, IL-5, and IL-4 in healthy (control), sensitized (OVA), and OVA + YM976-treated guinea pigs

Group	PAF plasma (pg/ml)	IL-5 plasma (pg/ml)	IL-4 plasma (pg/ml)
Control	417.40 ± 17.80	0.021 ± 0.003	0.023 ± 0.002
OVA	465.10 ± 11.20*	0.048 ± 0.001***	0.053 ± 0.015*
OVA + YM976	437.10 ± 11.90	0.046 ± 0.003	0.028 ± 0.004+

Data are means ± SE
PAF platelet activating factor, *IL* interleukin
*$p < 0.001$ vs. control, ***$p < 0.001$ vs control, +$p < 0.05$ vs. ovalbumin

(Mokrý et al. 2013; Christensen et al. 2012; Page and Spina 2012).

PDE4 inhibitors of the second generation (piclamilast, cilomilast, orroflumilast) are clinically useful as bronchodilating, anti-inflammatory, and antitussive agents; the effects being also confirmed in experimental conditions. These findings emphasize the potential advantage that could be gained from testing other PDE4 inhibitors (Fabbri et al. 2010; Giembycz and Field 2010). In the present study we tested the YM976, a PDE4 inhibitor, at a dose of 1.0 mg/kg. This dose was chosen based on the previous experiments performed in different animal species such as the rat, the mouse, the ferret, and the guinea pig. In guinea pigs, oral doses of YM976 of 1.0 and 0.5 mg/kg are equivalent to ED_{50} for inhibition of eosinophil infiltration and airway hyperreactivity, respectively (Aoki et al. 2001b). The present study demonstrates that YM976 in a dose of 1.0 mg/kg, given intraperitoneally, appreciably suppressed cough and *in vivo* airway reactivity after both single and repetitive several day long administration. These results are in line with our previous experiments, in which a significant antitussive effect of PDE4 inhibition with citalopram had been demonstrated after a single 1.0 mg/kg dose (Mokrý et al. 2008). In contrast, rolipram, at the same dose, produced no significant antitussive effect (Mokrý et al. 2013). The discrepancy can be partly explained by a different affinity of these drugs to the PDE4B isoform which has to do with the anti-inflammatory and bronchodilating effects. Other studies have demonstrated only a mild emetogenic activity of YM976, with no emesis observed at the oral dose of 10.0 mg/kg in ferrets, which suggests a relatively safe profile of this PDE4 inhibitor compared to emetogenic propensity of rolipram or roflumilast in asthmatic patients (Gupta 2012; Aoki et al. 2001a; Aoki et al. 2000b).

The antitussive effect observed in the present study was associated with a decrease in airway resistance, in which the involvement of smooth muscle relaxation and subsiding mucosal edema and secretion might be presupposed. Suppression of tracheal strips contractility in response to acetylcholine and histamine suggests a link between airway smooth muscle relaxation and antitussive effect. We failed to find suppression of *in vitro* airway reactivity in lung strips. That could be explained by different lung tissue structure, where there is a substantial presence of vascular smooth muscles, with lower cAMP and higher cGMP-coupled relaxation. As YM976 is highly selective for PDE4 that is responsible for cAMP degradation, its inhibition should not affect the cytoplasmic level of cGMP; thus, the relaxing process of vascular smooth muscle is negligible. Similar data on *in vitro* lung tissue reactivity have been recently demonstrated with rolipram, where no significant differences between the rolipram-treated and placebo-treated groups were observed in the ovalbumin-sensitized guinea pigs (Mokrý et al. 2013). In contrast, another PDE4 inhibitor, citalopram, appreciably suppresses airway reactivity, in both lung and trachea, in *in vitro*, but not *in vivo,* condition (Mokrý et al. 2008). Further, Moriuchi et al. (2003) have demonstrated a potent relaxing effect of YM976 and rolipram in bovine tracheal smooth muscle contracted by histamine, but not methacholine. Both these PDE4 inhibitors were more potent compared to theophylline, and they evoked an increase in cAMP in tracheal smooth muscle cells. However, YM976 inhibits the inflammatory cell function more intensely than rolipram does (Aoki et al. 2000a). That suggests that YM976 may suppress antigen-induced bronchoconstriction *via* inhibition of mediator release from leukocytes, and not by a direct relaxation effect (Moriuchi et al. 2003).

Selective inhibitors of PDE4, such as rolipram, citalopram, cilomilast, roflumilast, and piclamilast, have predominantly anti-inflammatory and immunomodulation effects, with much reduced adverse effects compared to non-selective ones (Karish and Gagnon 2006). In the present study, a significant decrease in specific airway resistance, a marker of *in vivo* airway reactivity, in the ovalbumin-sensitized and challenged guinea pigs was observed, suggesting a major role of YM976 in suppressing inflammation. A decrease in the accumulation of eosinophils induced by 7-day long administration of YM976 in both blood and BAL fluid supports that notion. Similar results have also been found in a study by Aoki et al. (2001b), where oral administration of YM976 in the ovalbumin-sensitized guinea pigs dampened the antigen-induced airway responses due to anti-inflammatory action, rather than direct tracheal relaxation.asIn that study, YM976 also decreased plasma-leakage and airway eosinophil infiltration. In line with the findings outlined above, YM976 suppresses the accumulation of eosinophils in rats and IL-5 production in a mouse model of ovalbumin-induced eosinophilia (Aoki et al. 2000a). In the present model of ovalbumin-sensitized guinea pigs, the levels of PAF, a potent stimulus for eosinophils accumulation in asthma, and IL-4 and IL-5, released from Th2 cells and stimulating eosinophils and mast cells *via* eotaxin expression in asthma, were significantly increased. These results give assurance to the validity of sensitization. However, only IL-4 was significantly decreased in the YM976-treated guinea pigs, which is explicable by a relatively short-term, 7 days, administration of the inhibitor, whereas the anti-inflammatory effects usually call for a longer administration.

In this study, we focused on the effects of repetitive doses of YM976. We failed to note significant differences in the number of coughs and in *in vivo* airway reactivity between single and repetitive doses of YM976. The *in vitro* data and inflammatory cell counts were assessed only after 7 days of YM976 administration. In a study of Aoki et al. (2000a) in the rat model of eosinophilia induced by repeated antigen exposure, a significant decrease in ED_{50} necessary to suppress eosinophilia was observed depending on single or chronic YM976 administration (1.4 mg/kg *vs.* 0.32 mg/kg *per os*, respectively). In that study, however, ovalbumin and *Bordetella pertussis* were used as antigens during 22-day long sensitization and YM976 therapy started from the first day of sensitization.

There have so far been no relevant data about antitussive effects of YM976. In this regard, our results indicate a potential efficacy of YM976 in suppressing cough. The model used in our experiments is considered as a partial model of bronchial asthma, with the predominance of eosinophils infiltration (Mokrý et al. 2008). Other models, e.g., COPD models with the predominance of neutrophils, should be tested to exclude or confirm the efficacy of PDE4 inhibitors for cough suppression.

5 Conclusions

We believe we have demonstrated that the selective PDE4 inhibitor YM976 suppresses citric-acid induced cough, airway reactivity, and markers of inflammation in ovalbumin-sensitized guinea pigs. The YM976 anti-inflammatory, smooth muscle relaxing, and cough suppressing influence might be of potential benefit in disorders underlain by allergic eosinophilic inflammation and cough.

Acknowledgements We thank K. Jesenska for technical assistance. This study was supported in parts by VEGA grant 1/0260/14, Slovak Research and Development

Agency under the contract no. APVV-0305-12, grant MZ2012/35-UKMA-12, and by the project "Biomedical Center Martin" ITMS code: 26220220187, the project is co-financed from EU sources.

Conflicts of Interest The authors declare no conflicts of interest in relation to this article.

References

Abbott-Banner KH, Page CP (2014) Dual PDE3/4 and PDE4 inhibitors: novel treatments for COPD and other inflammatory airway diseases. Basic Clin Pharmacol Toxicol 114:365–376

Aoki M, Fukunaga M, Kitagawa M, Hayashi K, Morokata T, Ishikawa G, Kubo S, Yamada T (2000a) Effect of a novel anti-inflammatory compound, YM976, on antigen-induced eosinophil infiltration into the lungs in rats, mice, and ferrets. J Pharmacol Exp Ther 295:1149–1155

Aoki M, Kobayashi M, Ishikawa J, Saita Y, Terai Y, Takayama K, Miyata K, Yamada T (2000b) A novel phosphodiesterase type 4 inhibitor, YM976 (4-(3-chlorophenyl)-1,7-diethylpyrido[2,3-d] pyrimidin-2(1H)-one), with little emetogenic activity. J Pharmacol Exp Ther 295:255–260

Aoki M, Fukunaga M, Sugimoto T, Hirano Y, Kobayashi M, Honda K, Yamada T (2001a) Studies on mechanisms of low emetogenicity of YM976, a novel phosphodiesterase type 4 inhibitor. J Pharmacol Exp Ther 298:1142–1149

Aoki M, Yamamoto S, Kobayashi M, Ohga K, Kanoh H, Miyata K, Honda K, Yamada T (2001b) Antiasthmatic effect of YM976, a novel PDE4 inhibitor, in guinea pigs. J Pharmacol Exp Ther 297:165–173

Bender AT, Beavo JA (2006) Cyclic nucleotide phosphodiesterases: molecular regulation to clinical use. Pharmacol Rev 58:488–520

Christensen I, Miskovicova H, Porvaznik I, Joskova M, Mokra D, Mokry J (2012) Selective inhibition of phosphodiesterase 7 (PDE7) by BRL50481 in healthy and ovalbumin-sensitized guinea pigs. Acta Medica Martiniana 12:16–23

Fabbri LM, Beghe B, Yasothan U, Kirkpatrick P (2010) Roflumilast. Nat Rev Drug Discov 9:761–762

Franová S, Nosalova G, Pechanova O, Sutovska M (2007) Red wine polyphenolic compounds inhibit tracheal smooth muscle contraction during allergen-induced hyperreactivity of the airways. J Physiol Pharmacol 59:727–732

Fujimura M, Liu Q (2007) Selective inhibitors for phosphodiesterase 3 and 4 in antigen-induced increase of cough reflex sensitivity in guinea pigs. Pulm Pharmacol Ther 20:543–548

Giembycz MA, Field SK (2010) Roflumilast: first phosphodiesterase 4 inhibitor approved for treatment of COPD. J Drug Des Devel Ther 4:147–158

Gupta S (2012) Side-effects of roflumilast. Lancet 379:710–711

Karish SB, Gagnon JM (2006) The potential role of roflumilast: the new phosphodiesterase-4 inhibitor. Ann Pharmacother 40:1096–1104

Kitta T, Tanaka H, Mitsui T, Moriya K, Nonomura K (2008) Type 4 phosphodiesterase inhibitor suppresses experimental bladder inflammation. Br J Urol Int 102:1472–1476

Medvedová I, Prso M, Eichlerova A, Mokra D, Mikolka P, Mokry J (2015) Influence of roflumilast on airway reactivity and apoptosis in ovalbumin-sensitized guinea pigs. Adv Exp Med Biol 838:11–18

Mokrý J, Mokra D (2013) Immunological aspects of phosphodiesterase inhibition in the respiratory system. Respir Physiol Neurobiol 187:11–17

Mokrý J, Nosalova G (2011) The influence of the PDE inhibitors on cough reflex in guinea pigs. Bratisl Lek Listy 112:131–135

Mokrý J, Mokra D, Nosalova G, Beharkova M, Feherova Z (2008) Influence of selective inhibitors of phosphodiesterase 3 and 4 on cough and airway reactivity. J Physiol Pharmacol 59:473–482

Mokrý J, Nosalova G, Mokra D (2009) Influence of xanthine derivatives on cough and airway reactivity in guinea pigs. J Physiol Pharmacol 60:87–91

Mokrý J, Joskova M, Mokra D, Christensen I, Nosalova G (2013) Effects of selective inhibition of PDE4 and PDE7 on airway reactivity and cough in healthy and ovalbumin-sensitized guinea pigs. Adv Exp Med Biol 756:57–64

Moriuchi H, Nakahara T, Maruko T, Sakamoto K, Ishii K (2003) Relaxant effect of YM976, a novel phosphodiesterase 4 inhibitor, on bovine tracheal smooth muscle. Eur J Pharmacol 470:57–64

Morjaria JB, Dickinson RS, Morice AH (2013) Novel antitussive strategies. Drug Discov Today 18:380–388

Nosalová G, Mokrý J (2001) The mechanism of action of xanthine derivatives and suppression of cough. Acta Medica Martiniana 1:14–18

Page CP, Spina D (2012) Selective PDE inhibitors as novel treatments for respiratory diseases. Curr Opin Pharmacol 12:275–286

Sutovska M, Franová S, Sutovsky J (2009) The influence of animal species on the relationship between ATP-sensitive potassium ion channels and defense reflexes of the airways. Bratisl Lek Listy 110:269–275

Advs Exp. Medicine, Biology - Neuroscience and Respiration (2016) 24: 71–80
DOI 10.1007/5584_2016_244

Published online: 10 May 2016

Airway Defense Control Mediated *via* Voltage-Gated Sodium Channels

M. Kocmalova, M. Joskova, S. Franova, P. Banovcin, and M. Sutovska

Abstract

Expression of voltage-gated sodium channels (Na_v) takes place in the airways and the role of $Na_v1.7$ and $Na_v1.8$ in the control of airway's defense reflexes has been confirmed. The activation of Na_v channels is crucial for cough initiation and airway smooth muscle reactivity, but it is unknown whether these channels regulate ciliary beating. This study evaluated the involvement of $Na_v1.7$ and $Na_v1.8$ channels in the airway defense mechanisms using their pharmacological blockers in healthy guinea pigs and in the experimental allergic asthma model. Asthma was modeled by ovalbumin sensitization over a period of 21 days. Blockade of $Na_v1.7$ channels significantly decreased airway smooth muscle reactivity *in vivo*, the number of cough efforts, and the cilia beat frequency in healthy animals. In the allergic asthma model, blockade of $Na_v1.8$ efficiently relieved symptoms of asthma, without adversely affecting cilia beat frequency. The study demonstrates that $Na_v1.8$ channel antagonism has a potential to alleviate cough and bronchial hyperreactivity in asthma.

Keywords

Airways • Asthma • Bronchial hyperreactivity • Ciliary beat frequency • Cough • Voltage-gated sodium channels

M. Kocmalova, M. Joskova (✉), S. Franova, and M. Sutovska
Department of Pharmacology, Division of Respirology, BioMed Martin, Jessenius Faculty of Medicine, Comenius University in Bratislava, 4C Mala Hora St, 036 01 Martin, Slovakia
e-mail: Michaela.Kocmalova@jfmed.uniba.sk; joskova@jfmed.uniba.sk

P. Banovcin
Department of Children and Adolescents, Jessenius Faculty of Medicine in Martin, Comenius University in Bratislava, and Martin University Hospital, Martin, Slovakia

1 Introduction

Motor and sensory innervation of airways and lungs is mediated by the vagal nerves, the dysfunction or dysregulation of which contributes to the pathogenesis of asthma and chronic obstructive pulmonary diseases with bronchial smooth muscle contraction, cough, and a greater mucus secretion (Undem and Kollarik 2005).

Bronchoconstriction and cough are important airway defense mechanisms of reflex origin that protect the lungs from inspiring noxious agents. Bronchoconstriction can be evoked from the whole tracheobronchial tree, while the cough reflex is triggered from the larynx and proximal airway branches (Karlsson et al. 1988).

The vagal control of mucociliary clearance, the major components of which include the sol (periciliary layer), gel (mucus layer), and the cilia embedded in the periciliary fluid has been debated. Kollarik et al. (2003) have shown that the bronchial sensory nociceptive C-fibers innervate the mucosa of mice and humans. Other studies provide evidence that ciliary beat frequency (CBF) and mucus secretion are vagally mediated processes (Undem and Kollarik 2005; Eljamal et al. 1985).

Peripheral stimuli for airway reflexes interact with afferent nerve terminals through various receptors and ion channels to cause membrane depolarization. This, in turn, activates the voltage-gated sodium channels (Na_V) that are responsible for action potential generation and relaying the signal to the brainstem (Muroi and Undem 2014). Na_v channels are transmembrane proteins consisting of the main α (channel pore) and complementary β subunits. Nine subtypes of Na_v ion channels, marked $Na_v1.1$–$Na_v1.9$, have by far been identified (Vetter et al. 2012). Each of the Na_v channel subtype has unique kinetic properties that differentiate its effect on electrogenesis and propagation of action potentials from the other channels. The overall voltage-gated sodium current is a doubly biphasic process in sensory neurons; rapid current activation/deactivation followed by a slow similarly biphasic process. The kinetics of the currents and their sensitivity to tetrodotoxin (TTX) help differentiate the relative role of sodium channel subtypes. Thus, the fast component of Na_v current is blocked by TTX, whereas the slow component is TTX resistant due to the presence of the TTX-resistant channels $Na_V1.8$ and 1.9 (Muroi et al. 2011).

$Na_v1.7$, $Na_v1.8$, and $Na_v1.9$, and less $Na_v1.2$ and $Na_v1.3$ channels are expressed in vagal afferent nerve fibers and C-fibers innervating airways. The first three channels outlined above are upregulated by inflammation (Muroi and Undem 2011). Na_v channels have also been identified in several types of isolated smooth muscle cells (vascular, urinal, and gastrointestinal) and in cultured smooth muscle cells (bronchial, coronary arterial, pulmonary arterial, and aortic) (Jo et al. 2004). Lidocaine and other local anesthetics can block Na_v channels in afferent vagal nerves, which also inhibits the cough reflex. However, use of selective $Na_v1.7$, $Na_v1.8$, or $Na_v1.9$ blockers seems to have a more direct and devoid of side effects antitussive action (Muroi and Undem 2011). $Na_v1.7$ channels are mostly present in vagal sensory C-fibers and Aδ-fibers in the airways (Kwong et al. 2008). Cummins et al. (1998) have demonstrated that opening of $Na_v1.7$ channle is dependent on the generation of action potential. This channel plays a crucial role in the regulation of neuronal excitability and action potential conduction in the majority of vagal afferent nerves. The $Na_v1.8$ channel is TTX-resistant and its distinguishing feature from the $Na_v1.7$ and other subtypes is rapid activation and deactivation dependent on voltage (Muroi and Undem 2011). Another important feature of this channel is a high threshold for activation, which is significantly reduced after exposure to inflammatory mediators. The $Na_v1.8$ channel currents are also involved in the prostaglandin E2-evoked sensitization of pulmonary C neurones during airway inflammation (Kwong and Lee 2005).

Recently several preferential, small molecule Na_v blockers have been developed (Priest 2009). It is difficult to propose a uniquely selective blocker due to high channels homologicity. Side effects on the heart, central nervous system, and skeletal muscles after systemic administration of blockers result from the ubiquitous expression of these channels. In the present study we set out to evaluate the role of $Na_v1.7$ and $Na_v1.8$ channels in the airway defense mechanisms during physiological and pathological conditions. We addressed the issue by examining the cough reflex and specific airway resistance *in vivo*, and ciliary beat frequency (CBF) *in vitro*, the features that reflect bronchial reactivity and mucociliary clearance.

2 Methods

The study was approved by the Ethics Committee of the Jessenius Faculty of Medicine in Martin accordance with the revised Declaration of Helsinki of 1998. The experiments followed the EU criteria of experimental animal well-fare (decision No. EK 1249/2013 and 1386/2013). The animals used in the study were adult male TRIK strain guinea pigs weighing 150–350 g. They were obtained from the Department of Experimental Pharmacology, Slovak Academy of Sciences, Dobra Voda in Slovakia and from the breeding facility Velaz, Prague in Czech Republic, and were housed in an approved animal holding facility. Animals were adapted to laboratory conditions before the start of experiment: ambient temperature of 21–24 °C, relative humidity of 55 ± 10 %, and air conditioned atmosphere.

2.1 Study Design

A total of 140 animals were used in this study. They were divided into 14 groups of ten animals each. Experimental airway inflammation was modeled in seven groups. The sensitization procedure consisted of repetitive parenteral administration of the allergen ovalbumin (OVA), adsorbed on aluminium hydroxide (5 mg OVA and 100 mg $Al(OH)_3$). Briefly, the allergen was administered doubly, intraperitoneally and subcutaneously, each 0.5 ml, on Day 1, then only i.p. on Day 4, and only s.c. on Day 9. In addition, allergen was administered also by inhalation during on Days 12, 15, 18, and 20 by means of using a double-chamber body plethysmograph box (HSE type 855; Hugo Sachs Elektronik, March, Germany).

Sensitization was followed by a number of acute pharmacological tests. Pharmacological agents were used in both *in vivo* and *in vitro* conditions and were executed 24 hours after the last OVA inhalation challenge. The following agents were used *in vivo:* carbamazepine – $Na_v1.7$ sodium channel blocker, A803467 – $Na_v1.8$ blocker, salbutamol, and codeine, and *in vitro*: dimethyl sulfoxide (DMSO), A803467, NAV26 – selective $Na_v1.7$ channel blocker, and licarbazepine. The substances NAV26 and licarbazepine were used only *in vitro,* because of their pharmacological properties; NAV26 also blocks $Na_v1.5$ channels and causes serious cardiac side effect *in vivo* and licarbazepine is an active metabolite of carbamazepine. All chemical agents used are listed in Table 1.

The course of experiments:

In vivo:

- Healthy (OVA-) and sensitized (OVA+) animals – administration of vehicles;
- OVA- and OVA+ animals – carbamazepine, 10 mg/kg i.p.;
- OVA- and OVA+ animals – A803467, 1 mg/kg i.p.;
- OVA- and OVA+ animals – salbutamol, 10 mg/kg i.p.;
- OVA- and OVA+ animals – codeine, 10 mg/kg per os.

In vitro – local application directly on biological samples obtained from animals:

- OVA- and OVA + – vehicles
- OVA- and OVA + – licarbazepine, concentrations 10^{-7}, 10^{-6}, 10^{-5} mol/L;
- OVA- and OVA + – A803467, concentrations 10^{-7}, 10^{-6}, 10^{-5} mol/L;
- OVA- and OVA + – NAV26, concentrations 10^{-7}, 10^{-6}, 10^{-5} mol/L;
- OVA- and OVA+ 10 % DMSO.

2.2 Cough Reflex Assessment

Cough reflex was assessed by inhalation of aerosol of 0.3 mol/L of citric acid (CA) in saline. The aerosol was generated by a PARI jet nebulizer (Paul Ritzau, Pari-Werk GmbH, Starnberg,

Germany, output 5 L/s, particles mass median diameter 1.2 μm) and was delivered into the nasal chamber of the body box for 3 min. Two trained observes counted the number of cough efforts. The cough response was later re-evaluated and verified against the computer records of airflow curve changes in the nasal body box chamber. The cough response was assessed prior to (baseline measurement, N value in figures) and then 1, 2, and 5 h after administration of vehicles, codeine, and the Na_v blockers tested. A 2-h time interval was observed between the use of each agent.

2.3 Evaluation of Airway Smooth Muscle Reactivity *in Vivo*

Specific airway resistance (sRaw) was taken as a measure of airway smooth muscle reactivity *in vivo*. The awake guinea pig was placed in a double-chamber body box (HSE type 855; Hugo Sachs Elektronik, March, Germany) consisting of nasal and body chambers. Changes in sRaw were recorded in response to a 1-minute inhalation of citric acid and the bronchoconstrictors histamine (10^{-6} mol/L) and metacholine (10^{-6} mol/L) before (N value in figures) and 1, 2, and 5 h after application of these mediators as outlined in Table 1. There was 1 min interval between each bronchoconstrictor agent used during which fresh air was blown into nasal chamber. Airway reactivity in response to saline nebulization was used as reference. sRaw, expressed in $mL \cdot s^{-1}$ was calculated by the Pennock et al. (1979) method.

2.4 Ciliary Beat Frequency (CBF)

A sample of ciliated epithelium was brushed out from the trachea in the experimental *in vitro* conditions. Cytology brush (2.5 mm in diameter) was soaked in physiological saline before brushing, then inserted into the trachea and rotated gently to collect ciliated cells. Cilia samples were suspended in 1 ml of RPMI 1640 Medium (ThermoFisher Scientific; Waltham, MA), with temperature controlled at 37–38 °C (Temp Controller 2000-2; PeCon GmbH, Erbach, Germany) and microscopic glass slides were prepared. An inverted phase contrast microscope (Zeiss Axio Vert. A1; Carl Zeiss AG, Göttingen, Germany) was used to examine specimens. Damaged ciliated cells were excluded from the examination. Time series analysis of cilia beating frequency consisted of sequential 10-sec video recordings at 1-min intervals for 15 min by a digital camera (Basler A504kc; Basler AG, Ahrensburg, Germany) at a frame rate from 256 to 512 per sec. The recordings were assessed by Ciliary Analysis

Table 1 Specifications of used chemicals

Agent	Company	Solvent
Salbutamol	Sigma Aldrich (SR)	aqua pro injectione
Carbamazepine	Sigma Aldrich (SR)	aqua pro injectione (10 % solubility)
OVA	Sigma Aldrich (SR)	0.9 % NaCl
Histamine	Sigma Aldrich (SR)	0.9 % NaCl
DMSO solution	Sigma Aldrich (SR)	–
A803467	TOCRIS (USA)	10 % DMSO
Licarbazepine	TOCRIS (USA)	10 % DMSO
NAV26	TOCRIS (USA)	10 % DMSO
Codeine	Slovakofarma Hlohovec (SR)	aqua pro injectione
Sodium chloride solution	ApliChem (Germany)	–
Metacholine	ApliChem (Germany)	0.9 % NaCl
RPMI 1640 medium	Invitrogen/Gibco (USA)	–
$Al(OH)_3$	CentralChem (SR)	0.9 % NaCl
Citric acid	ACROSorganics (SR)	0.9 % NaCl

software (LabVIEW™) to generate a ciliary region of interest. The assessment method was based on the frequency analysis of the intensity variance curve, implementing the Fast Fourier Transform algorithm (Hargaš et al. 2011). Potential artifacts were filtered out by comparison of the ciliary region of interest with a relevant video recording. CBF was taken as a kinematic parameter for describing the ciliary movement in the airways.

2.5 Statistical Analysis

Data are given as means ± SE. Statistical significance of differences was assessed using one-way ANOVA with the Bonferroni *post-hoc* test. A p-value <0.05 defined significant changes.

3 Results

3.1 Chemically Induced Cough

Carbamazepine was effective in decreasing the number of cough efforts in the physiological condition, i.e., OVA- group, at each of the assessed time intervals after its administration (Fig. 1a). The selective $Na_v1.8$ sodium channel blocker A803467 was here less effective as it decreased the number of coughs 5 h after administration only. In contrast, in the allergic condition, i.e., OVA+ group, both carbamazepine and A803467 significantly reduced the number of coughs starting from the first hour after administration on (Fig. 1b). The effect of A803467 was here akin to that of codeine, the archetype antitussive agent.

3.2 Airway Smooth Muscle Reactivity (ASM) *in Vivo*

In addition to the antitussive ability of the selective $Na_v1.8$ sodium channel blocker A803467, this substance significantly decreased the sRaw response to airway irritation by the different bronchoconstrictors used in the OVA+ animals. The blocker was not as effective in the OVA- group of non-sensitized, healthy animals. Carbamazepine did not significantly affect sRaw in the OVA+ animals, even with a tendency for increased contractile airway activity in the tachykinin-mediated bronchoconstriction. On the other side, sRaw values were significantly decreased by carbamazepine in the OVA- healthy group (Table 2).

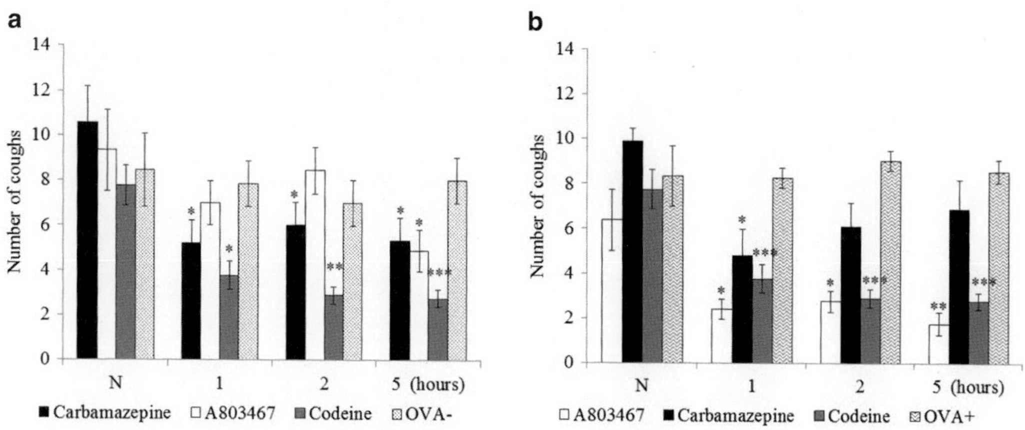

Fig. 1 The number of cough efforts after administration of $Na_v1.7$ and $Na_v1.8$ blockers in the physiological OVA- (**a**) and allergic OVA+ (**b**) conditions. N – baseline measurement before drugs administration; 1, 2, and 5 on the horizontal axes correspond to the consecutive hours of cough assessment. OVA- and OVA+ bars are internal controls in each panel, depicting the lack of changes in cough with elapsing time without the use of any pharmacological tools. $^*p < 0.05$, $^{**}p < 0.01$, and $^{***}p < 0.001$ *vs*. N

Table 2 Changes in specific airway resistance (sRaw in mL · s^{-1}) in response to bronchoconstricting mediators before (N) and after application of $Na_v1.7$ and $Na_v1.8$ blockers in the physiological OVA- and allergic OVA+ conditions

	After citric acid				After histamine				After methacholine				
	N	1 h	2 h	5 h	N	1 h	2 h	5 h	N	1 h	2 h	5 h	
Salbutamol	11.9 ± 5.5	7.8 ± 1.0	7.4 ± 1.2	9.5 ± 1.3	4.5 ± 2.6	2.0 ± 0.7*	5.4 ± 1.4	4.8 ± 0.9	4.7 ± 3.5	1.9 ± 1.9*	5.5 ± 1.6	4.8 ± 2.7	OVA-
	9.3 ± 1.6	8.8 ± 1.6	7.2 ± 2.0	8.0 ± 1.4	10.8 ± 0.8	7.6 ± 0.9**	7.8 ± 1.7	11.5 ± 1.2	18.9 ± 7.1	8.0 ± 3.0**	8.6 ± 3.3**	9.6 ± 3.6**	OVA+
Carbamazepine	7.3 ± 2.5	6.3 ± 2.3	4.5 ± 1.3*	5.9 ± 1.0	11.6 ± 1.9	7.4 ± 2.0**	6.7 ± 2.0	14.5 ± 2.5	20.7 ± 3.6	6.7 ± 1.8*	12.2 ± 4.2	14.5 ± 2.5	OVA-
	9.1 ± 2.9	9.8 ± 4.5	9.8 ± 2.3	11.7 ± 4.1	15.9 ± 4.3	9.0 ± 3.3	14.0 ± 4.2	12.6 ± 2.9	15.5 ± 15.9	16.5 ± 5.6	15.7 ± 4.4	19.7 ± 5.0	OVA+
A803467	12.3 ± 2.9	3.9 ± 0.6*	6.7 ± 2.1	10.1 ± 1.8	14.8 ± 3.3	7.9 ± 1.9	12.5 ± 3.4	14.1 ± 3.5	18.3 ± 3.7	10.3 ± 2.6	10.2 ± 2.5	11.9 ± 2.3	OVA-
	23.1 ± 4.9	6.3 ± 0.8***	11.6 ± 1.9***	10.15 ± 2.6***	28.2 ± 1.4	12.8 ± 3.0***	19.8 ± 3.3*	14.5 ± 2.7***	34.4 ± 3.9	14.3 ± 2.6***	18.2 ± 2.4**	17.3 ± 2.9***	OVA+

OVA ovalbumin, *N* baseline measurement before drug application; 1, 2, and 5 – consecutive hours of measurements after drug application

*p < 0.05, **p < 0.01, and ***p < 0.001 *vs.* N

3.3 Ciliary Beat Frequency (CBF)

CBF was significantly decreased in the OVA+ sensitized animals with allergic inflammation. The Na_v channel blockers licarbazepine, NAV26, and A803467 reduced CBF in both non-sensitized OVA- (Figs. 2a, 3a, and 4a) and sensitized OVA+ groups (Figs. 2b, 3b, and 4b), although the decrease was insignificant in the latter group.

CBF was notably decreased at 10^{-6} and 10^{-5} mol/L concentrations of licarbazepine and NAV26 in the OVA- groups (Figs. 2a and 3a). The $Na_v1.8$ blocker A803467 reduced CBF at the smallest 10^{-7} mol/L concentration in the non-sensitized animals. However, it tended to have an opposite stimulating effect at higher concentrations (Fig. 4a). We also followed changes in CBF after application of the solvent DMSO during physiological and pathological inflammatory conditions with no observable negative effect.

4 Discussion

This study was focused on the role of voltage-gated sodium channels in the airway defense mechanisms. We found that blockers of $Na_v1.7$ and $Na_v1.8$ channels decrease cough and airway resistance *in vivo* and reduce ciliary beat frequency in both healthy and allergic asthma-like conditions in guinea pigs. These effects were notably expressed for the $Na_v1.7$ in the former condition and for $Na_v1.8$ in the latter pathological condition. The findings support the notion the channels play an essential role in the control of airway defense control.

Cough is typically thought of as a reflex response, requiring primary afferent input to the

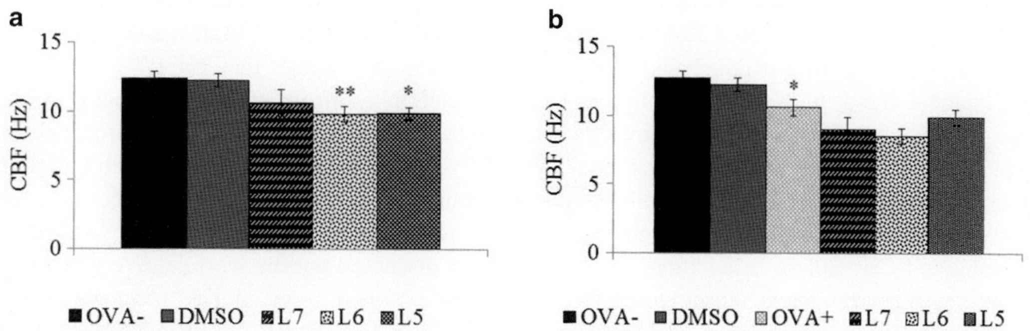

Fig. 2 Ciliary beat frequency (CBF) in non-sensitized OVA- (**a**) and sensitized OVA+ (**b**) animals after application of licarbazepine in the concentrations as follows: L7 – 10^{-7}, L6 – 10^{-6}, and L5 – 10^{-5} mol/L. *$p < 0.05$, **$p < 0.01$, and ***$p < 0.001$ *vs.* OVA-

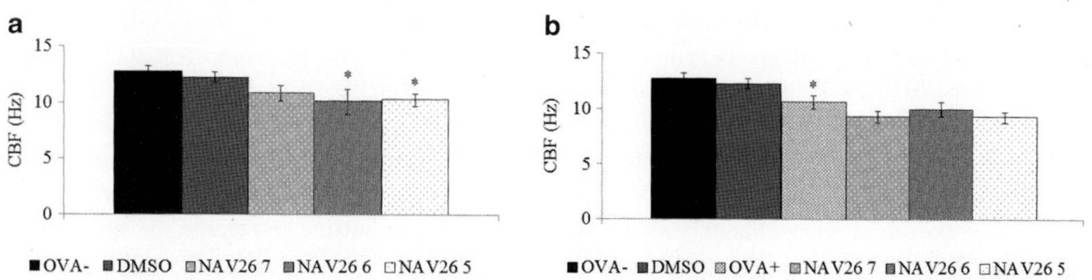

Fig. 3 Ciliary beat frequency (CBF) in non-sensitized OVA- (**a**) and sensitized (**b**) OVA+ animals after application of NAV26 in the concentrations as follows: NAV26 7 – 10^{-7}, NAV26 6 – 10^{-6}, and NAV26 5 – 10^{-5} mol/L. *$p < 0.05$, **$p < 0.01$, and ***$p < 0.001$ *vs.* OVA-

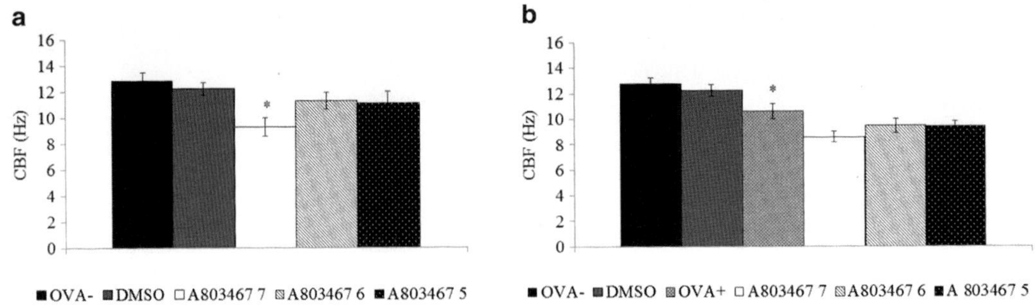

Fig. 4 Ciliary beat frequency (CBF) in non-sensitized OVA- (**a**) and sensitized (**b**) OVA+ animals after application of A803467 in the concentrations as follows: A803467 7 – 10^{-7}; A803467 6 – 10^{-6}; and A803467 5 – 10^{-5} mol/L; $^{*}p < 0.05$, $^{**}p < 0.01$, and $^{***}p < 0.001$ *vs.* OVA-

brainstem leading to reflex changes in respiration. The afferent innervation of airways and lungs responsible for mediating cough is derived from either the nodose or jugular vagal ganglia, from which the pulmonary chemosensor C- and Aδ-fibers arise. These fibers are largely quiescent in normal airways, but are recruited during airway irritation or inflammation (Narula et al. 2014). Lidocaine is used to inhibit cough reflex in bronchoscopy procedures indicating the involvement of Na_v in the reflex circuit. $Na_v1.8$, and $Na_v1.9$ channels have been shown to be expressed in the lung-specific vagal sensory neurons in experimental animal models that display a cough reflex such as the guinea pig (Kwong and Carr 2015). In addition, $Na_v1.7$ is expressed in the respiratory sensory vagal C- and Aδ-fibers (Muroi and Undem 2011; Kwong et al. 2008). A decrease in the number of coughs caused by a $Na_v1.8$ blocker in allergic inflammation found in the present study is in line with the results of Muroi and Undem (2011). That study reports up-regulation of $Na_v1.8$ channel expression and an upward shift of activation threshold after exposure of sensory neurons to inflammatory mediators. All these results raise a biological plausibility that $Na_v1.7$ and $Na_v1.8$ channels might become a target for the development of antitussive agents.

The expression of Na_v channels is a key feature for the initiation and conduction of action potentials in excitable tissues such as the heart, skeletal muscle and neurons. This includes the neurons that innervate the airway whose activity underlies respiratory sensations such as dyspnea and reflex changes including cough and bronchoconstriction (Kwong and Carr 2015). An airway stimulus, like allergen, can result in action potential generation. Subsequently, acetylcholine is released and activates M_3 receptors in airway smooth muscles, leading to bronchoconstriction (Undem and Carr 2010). The present findings that $Na_v1.7$ and $Na_v1.8$ channels play a role in airway smooth muscle contraction evoked by different bronchoconstricting mediators in both healthy and allergic conditions expand the knowledge on the voltage-gated sodium channels beyond their often discussed antitussive role. We also demonstrate a novel role of Na_v channels in the regulation of tracheal cilia. Blockade of these channels decreases ciliary beat frequency; the phenomenon being more pronounced in the healthy condition. Licarbazepine is an agent capable of inhibiting the $Na_v1.2$, $Na_v1.3$, and $Na_v1.7$ channels. However, a decrease in ciliary beat frequency by licarbazepine is due most like to its effect on the $Na_v1.7$, which may be inferred from a similar inhibitory effect of NAV26 that is over 85-fold more selective for $Na_v1.7$ than $Na_v1.5$ channels. Since Na_v channels are expressed in vagal afferent fibers of airways as outline above, the present findings lead support for the notion that CBF is under vagal control (Eljamal et al. 1985). We observed a different CBF response to Na_v blockers, depending on their selectivity and concentrations. Moderate and higher concentrations of non-selective $Na_V1.7$ blockers, NAV26 and licarbazepine were required for a stronger inhibitory effect on CBF, but a lower concentration of a selective $Na_V1.8$ blocker, A803467, sufficed to this end. There also may be an interaction between $Na_V1.8$ and transient receptor potential subfamily TRPV. High concentrations of the Na_V blockers used in the present study might cause an accumulation of second messengers acting in TRPV channel conduction with an inhibitory aftereffect on CBF (Joshi et al. 2009). A subtype of TRPV ion channels, known as TRPV4, is predominantly located in the tracheal epithelial cilia cells and participates in the regulation of CBF as well as mucociliary transport (Lorenzo et al. 2008). Airway mucus hypersecretion and increased mucus viscosity are typical asthma features. Mucociliary clearance is important for the prevention of mucus retention. However, antitussive drugs enhancing mucociliary transport should not affect CBF negatively, since that could increase the propensity for infection. It seems that Na_V blockers could fulfil such criteria for optimal antitussive preparations. Further research, using alternative study designs, is required to explorer this issue.

In conclusion, we demonstrate that both $Na_V1.7$ and $Na_V1.8$ play a key regulatory role in airway defense mechanisms such as hyperreactivity, cough reflex and mucociliary transport. In particular, selective blockers of $Na_V1.8$ channels are liable to have a good potential to alleviate cough and bronchial hyperreactivity in the allergic asthmatic condition, without having a detrimental effect on tracheal cilia function.

Acknowledgments This research was conducted within the project 'Measurement of Respiratory Epithelium Cilium Kinematics' and was supported by the grants VEGA 1/0165/14; MZ 2012/35-UK MA-12; APVV-0305-12; BioMed Martin (ITMS 26220220187); Center of Experimental and Clinical Respirology II, and 'The increasing of opportunities for career growth in research and development in the medical sciences', co-financed from EU sources. The authors would like to thank Katarina Jesenska for technical help during the experiments.

Conflicts of Interest The authors declare no conflict of interest in relation to this article.

References

Cummins TR, Howe JR, Waxman SG (1998) Slow closed-state inactivation: a novel mechanism underlying ramp currents in cells expressing the HNE/PN1 sodium channel. J Neurosci 18:9607–9619

Eljamal M, Wong LB, Yeates DB (1985) Capsaicin-activated bronchial- and alveolar-initiated pathways regulating tracheal ciliary beat frequency. J Appl Physiol 77:1239–1245

Hargaš L, Koniar D, Štofan S (2011) Sophisticated biomedical tissue measurement using image analysis and virtual instrumentation, practical applications and solutions using LabVIEW™ software, Dr. Silviu Folea (ed) ISBN: 978-953-307-650-8, InTech, Available from: http://www.intechopen.com/books/practical-applications-and-solutions-using-labview-software/sophisticatedbiomedical-tissue-measurement-using-image-analysis-and-virtual-instrumentation

Jo T, Nagata T, Iida H, Imuta H, Iwasawa K, Ma J, Hara K, Omata M, Nagai R, Takaziwa H, Nagase T, Nakajima T (2004) Voltage-gated sodium channel expressed in cultured human smooth muscle cells: involvement of SCN9A. FEBS Lett 567:339–343

Joshi SK, Honore P, Hernandez G, Schmidt R, Gomtsyan A, Scanio M, Kort M, Jarvis MF (2009) Additive antinociceptive effects of the selective $Na_V1.8$ blocker A803467 and selective TRPV1 antagonists in rat inflammatory and neuropathic pain models. J Pain 10:306–315

Karlsson JA, Sant'Ambrogio G, Widdicombe J (1988) Afferent neural pathways in cough and reflex bronchoconstriction. J Appl Physiol 65:1007–1023

Kollarik M, Dinh QT, Fischer A, Undem BJ (2003) Capsaicin-sensitive and -insensitive vagal bronchopulmonary C-fibers in the mouse. J Physiol 551:869–879

Kwong K, Carr MJ (2015) Voltage-gated sodium channels. Curr Opin Pharmacol 22:131–139

Kwong K, Lee LY (2005) Prostaglandin E2 potentiates a TTX resistant sodium current in rat capsaicin-sensitive vagal pulmonary sensory neurones. J Physiol 564:437–450

Kwong K, Carr MJ, Cibbard A, Savage TJ, Singh K, Jing J, Meeker S, Undem BJ (2008) Voltage-gated sodium channels in nociceptive versus non-nociceptive nodose vagal sensory neurons innervating Guinea pig lungs. J Physiol 586:1321–1336

Lorenzo IM, Liedtke W, Sanderson MJ, Valverde MA (2008) TRPV4 channel participates in receptor-operated calcium entry and ciliary beat frequency regulation in mouse airway epithelial cells. Proc Natl Acad Sci U S A 105:12611–12616

Muroi Y, Undem BJ (2011) Targeting peripheral afferent nerve terminals for cough and dyspnoe. Curr Opin Pharmacol 11:254–264

Muroi Y, Undem BJ (2014) Targeting voltage gated sodium channels NaV1.7, NaV1.8, and NaV1.9 for treatment of pathological cough. Lung 192:15–20

Muroi Y, Ru F, Kollarik M, Canning BJ, Hughes SA, Walsh S, Sigg M, Carr MJ, Undem BJ (2011) Selective silencing of NaV1.7 decreases excitability and conduction in vagal sensory neurons. J Physiol 589:5663–5676

Narula M, McGovern AE, Yang SK, Farrell MJ, Mazzone SB (2014) Afferent neural pathways mediating cough in animals and humans. J Thorac Dis 6:712–719

Pennock BE, Cox CP, Rogers RM, Cain WA, Wells JH (1979) A non-invasive technique for measurement of changes in specific airway resistance. J App Physiol 46:399–406

Priest BT (2009) Future potential and status of selective sodium channel blockers for the treatment of pain. Curr Opin Drug Discov Devel 12:682–692

Undem BJ, Carr MJ (2010) Targeting primary afferent nerves for novel antitussive therapy. Chest 137:177–184

Undem BJ, Kollarik M (2005) The role of vagal afferent nerves in chronic obstructive pulmonary disease. Proc Am Thorac Soc 2:355–360

Vetter I, Mozar CA, Durek T, Wingerd JS, Alewood PF, Christie MJ, Lewis RJ (2012) Characterisation of Na_v types endogenously expressed in human SH-SYSY neuroblastoma cells. Biochem Pharmacol 83:1562–1571

Advs Exp. Medicine, Biology - Neuroscience and Respiration (2016) 24: 81–85
DOI 10.1007/5584_2016_236

Published online: 02 May 2016

Virological Characteristics of the 2014/ 2015 Influenza Season Based on Molecular Analysis of Biological Material Derived from I-MOVE Study

E. Hallmann-Szelińska, K. Bednarska, M. Korczyńska, I. Paradowska-Stankiewicz, and L.B. Brydak

Abstract

The main goal of the international study I-MOVE (Influenza Monitoring of Vaccine Effectiveness) implemented in Poland is to identify and evaluate the activity types of influenza virus and to determine the effectiveness of vaccination against influenza in the 2014–2015 influenza season. The study is based on selecting patients with flu symptoms and collecting biological samples for laboratory examination. Detection, typing, and subtyping of influenza viruses were carried out by the National Center for Influenza Virus Research at National Institute of Public Health – National Institute of Hygiene, serving as a reference center, and also in selected laboratories of the Regional Sanitary Epidemiological Stations. Molecular biology methods, such as reverse transcription polymerase chain reaction (RT-PCR), were applied in this study. A total of 218 samples were collected. A hundred and twenty six samples, representing 57.8 % of the total, were confirmed with influenza virus infection. Influenza type A virus was detected in 54 samples, which included 16 samples of A/H1N1/pdm09 subtype and 11 samples of A/H3N2/ subtype. The remaining 27 samples positive for influenza type A were not subtyped. Influenza type B virus was detected in 57 samples, which appeared to be the dominant strain in this study. Furthermore, several cases of concurrent infection with influenza type B virus and the A/H1N/pdm09 or A/H3N2/ subtype were observed.

E. Hallmann-Szelińska (✉), K. Bednarska, and L.B. Brydak
Department of Influenza Research, National Influenza Center, National Institute of Public Health – National Institute of Hygiene, 24 Chocimska Street, 00-791 Warsaw, Poland
e-mail: ehallmann@pzh.gov.pl

M. Korczyńska and I. Paradowska-Stankiewicz
Department of Epidemiology, National Institute of Public Health – National Institute of Hygiene, 24 Chocimska Street, 00-791 Warsaw, Poland

Keywords
Influenza season • Influenza virus • Molecular biology • Vaccination • Virology • Virus subtype

1 Introduction

Since the time of Hong Kong pandemic in 1968–1969, circulation of two subtypes of influenza A virus and influenza B virus with a different intensity is recorded every year. In Poland, during several recent seasons influenza cases were registered from a few hundred to a few million suspected cases of influenza and influenza-like infections (ILI) (Bednarska et al. 2015; Czarkowski et al. 2014). Each season, influenza is identified by circulation of different subtypes of influenza type A virus. In the 2014/2015 season, the A/H3N2/ was dominant, while the major etiological agent of ILI was respiratory syncytial virus. In addition, since the 2011/2012 influenza season, various types of co-infection were recorded, both within the influenza viruses and viruses that cause ILI, regardless of patients' age. In the 2014–2015 influenza season, there was a three-component vaccine available; an inactivated split and subunit antigen of the following composition: A/California/7/2009 (H1N1) pdm09 – like virus, A/Texas/50/2012 (H3N2) – like virus, and B/Massachusetts/2012 – like virus (MMWR 2014). In Poland, the percentage of vaccinated population remains very low, as it amounted to a dismal 3.6 % in the season in question.

The international study I-MOVE (Influenza Monitoring of Vaccine Effectiveness) coordinated by the Epiconcept and European Center for Disease Prevention and Control (ECDC) has been conducted in European countries since 2007. Poland has entered the project as of the 2010/2011 influenza season (Kissling et al. 2012). For this case-controlled project, patients are qualified of every age who meet criteria of the ILI definition and from whom swabs are collected for laboratory examination conducted with reverse transcription polymerase chain reaction (RT-PCR) (Valenciano et al. 2015).

The aim of this article was to evaluate the activity of influenza virus and to determine the effectiveness of vaccination against influenza in the 2014–2015 season conducted within the framework of I-MOVE study in Poland. The control group included patients with ILI symptoms who had negative influenza test results.

2 Methods

The study was approved by an institutional Ethics Committee and was conducted in accord with the principles of the Declaration of Helsinki for Human Research. The study group consisted of 126 patients; 74 women and 52 men. The control group consisted of 91 patients, 46 women and 45 men, who visited the doctor's office for reasons other than influenza, who did not meet the ILI definition and had negative influenza RT-PCR results. The participants of both groups were stratified into four age-groups: 0–4, 5–14, 15–64, and 65+ years of age. The study material consisted of throat and nasal swabs obtained from patients who met the ILI criteria as set by the European Union. A total of 218 persons were enrolled into the study, recruited by 22 general practitioners (GP), collaborating in the study. Specimens were sent to the laboratory with attached information regarding the date of a visit to GP and swabbing. Specimens were stored at −80 °C until analysis.

2.1 RNA Isolation

Influenza virus RNA was isolated using a Maxwell 16 Viral Total Nucleic Acid Purification Kit (Promega Corporation, Madison, WI) from 200 μl of clinical samples suspended in phosphate-buffered saline, according to the manufacturer's instructions for low elution volume (LEV) cartridges. The RNA was eluted with 50 μl of RNase-free water.

2.2 Reverse Transcription PCR

Influenza A virus subtypes were determined using RT-PCR with a Light Thermocycler 2.0 System (Roche Diagnostics; Rotkreuz, Switzerland). RT-PCR reactions were conducted in capillary tubes of 20 μl volume using 0.5 μl (20 nM) of primers and 0.5 μl (5 nM) of probes for each reaction. Primers and probes were obtained through the Influenza Reagent Resource (IRR) program from the US Center for Disease Control (CDC) – Influenza Reagent Resource (IRR). The reaction mixture, containing reaction buffer, $MgSO_4$, bovine serum albumin (BSA), RNase-free H_2O, and SuperScript® III/Platinum® *Taq* Mix (Invitrogen by Life Technologies – Thermo Fisher Scientific, Carlsband, CA), was incubated with 5 μl of RNA sample in each capillary tube. RNA from the 2014/2015 vaccine viruses: A/California/7/2009(H1N1)pdm09 and A/Texas/50/2012 (H3N2), and B/Massachusetts/2/2012 were introduced as positive controls. The negative control consisted of RNase-free water. Before DNA amplification, the RNA templates were reverse transcribed to obtain the corresponding cDNA at 50 °C for 30 min. The cDNA was then subjected to initial denaturation (1 cycle of 95 °C for 2 min), followed by further amplification steps of denaturation at 95 °C for 15 s, annealing at 55 °C for 30 s, and elongation at 72 °C for 20 s repeated in 45 cycles.

3 Results and Discussion

Since the beginning of the influenza season in the middle of September 2014 (week 38 according to the ISO week date standard (ISO-8601)), Poland experienced only sporadic ILI cases. From week 43 to 47 of 2014, a widespread virus transmission began in the population and, consequently, we experienced a mild-to-moderate epidemic of influenza between week 48 of 2014 and week 15 of 2015, with the peak in weeks 10–13 of 2015. The influenza activity, which was mainly driven by influenza virus type A (51.4 %), subsided thereafter. There were no fatal cases of influenza recorded in the epidemic season of 2014–2015.

The majority of influenza cases taken into account in the present study concerned the after-peak period; still, however, with a substantially high, sustained level of transmission. Of the 218 cases that were medically attended, ILI was identified in 126 cases. In this group, there were only 36 (30.8 %) patients aged over 60 and eight of them tested positive for pandemic influenza A (H1N1). There were significant regional differences in the number of samples taken from patients who displayed the symptoms of ILI (Fig. 1). These differences distributed by the voivodeships of Poland are illustrated in Fig. 1. The differences lay down in the following way: Mazovian – 21, Lodz – 16, Silesian – 19, Subcarpathian – 62, Podlaskie – 21, Pomeranian – 20, Warmian-Masurian – 12, Lower Silesian – 31, and Greater Poland – 16 samples. Clinical material was tested to detect influenza virus type A and type B. Laboratory examination of the swabs collected from the Mazovian and Lodz provinces was performed in the Department of Influenza Research, National Influenza Center of the National Institute of Public Health-National Institute of Hygiene in Warsaw. Samples from other voivodeships were examined in the Provincial Sanitary-Epidemiological Stations. One hundred and twenty six samples, representing 57.8 % of the total, were confirmed with influenza virus infection, in the remaining 92 (42.2 %) samples influenza virus was not confirmed.

Influenza type A virus was detected in 54 samples, which included 16 samples of the subtype A/H1N1/pdm09 and 11 samples of the subtype A/H3N2/. The remaining 27 samples positive for influenza type A were not subtyped (Fig. 2).

In 57 samples influenza type B virus was detected, which appeared to be the dominant strain in this study. Only did five voivodeships confirm infections caused by influenza type B. The Subcarpathian voivodeship recorded 39 positive tests for influenza virus type B, which represented 68.4 % of all confirmed influenza virus type B

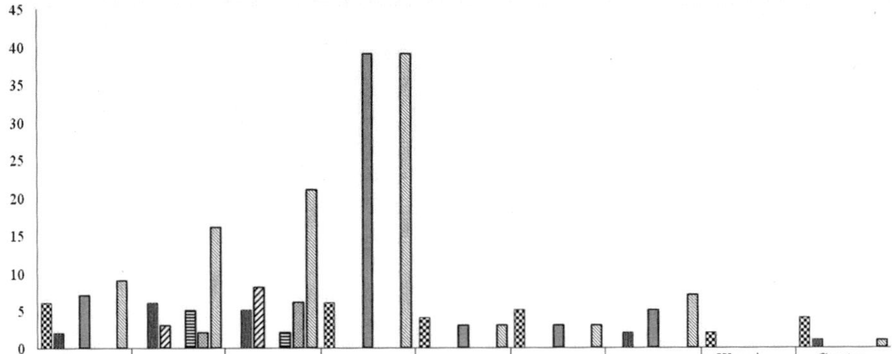

	Lower Silesian Voivodeship	Lodz Voivodeship	Masovian Voivodeship	Subcarpathian Voivodeship	Podlaskie Voivodeship	Pomeranian Voivodeship	Silesian Voivodeship	Warmian-Masurian Voivodeship	Greater Poland Voivodeship
Unsubtyped type A	6	0	0	6	4	5	0	2	4
A/H1N1/pdm09	2	6	5	0	0	0	2	0	1
A/H3N2/	0	3	8	0	0	0	0	0	0
Type B	7	0	0	39	3	3	5	0	0
A/H1N1/pdm09+ type B	0	5	2	0	0	0	0	0	0
A/H3N2/+ type B	0	2	6	0	0	0	0	0	0
Subtyped	9	16	21	39	3	3	7	0	1

Fig. 1 Virological characteristics of biological material collected within I-MOVE project in the epidemic season 2014–2015 distributed by the voivodships of Poland

Fig. 2 Virological characteristics of biological material collected within I-MOVE project in the epidemic season 2014/2015 in Poland

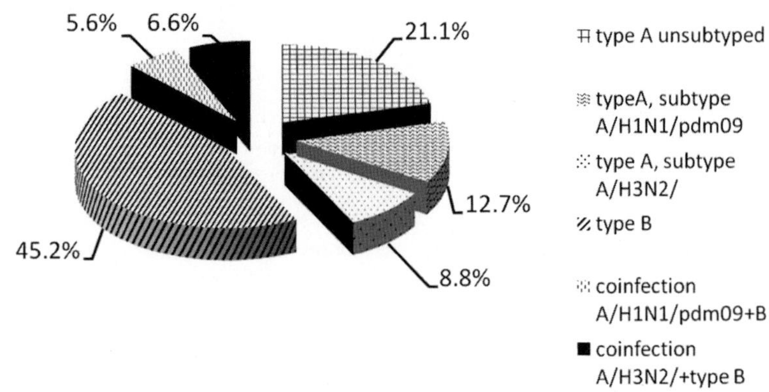

samples; type B virus infection constituted a fraction of that percentage in other voivodeships.

It is worthy of note that co-infections of influenza type B and either subtype A/H1N1/podm09 or subtype A/H3N2/ of influenza type A were also confirmed. Co-infections occurred in the Mazovian and Lodz voivodeships of Poland; 7 and 8 cases, respectively (Fig. 1 and Fig. 2). There was no association between the patient's age and the type of co-infection. In both voivodeships where the co-infections occurred, the age of the infected varied widely from 31 to 88 years.

Concerning the influenza type A virus, subtype A/H1N1/pdm09 predominated as it was detected in the following five voivodeships: Lodz, Mazovian, Lower Silesian, Silesian, and Greater Poland. That is in line with the data from the Global Influenza Surveillance and Response System and from the Department of Influenza Research, National Influenza Center of the National Institute of Public Health-National Institute of Hygiene in Warsaw, the organizations involved in the supervision of the 2014/2015 epidemic season, which confirmed the overall predominance of influenza virus type A, subtype A/H3N2/ in that season (Broberg et al. 2015).

4 Conclusions

- The main circulating strain of influenza virus in the 2014/2015 epidemic season in Poland was type A, subtype A/H3N2/, although subtype B was a strain frequently detected in some geographic regions.
- Substantial influenza morbidity in the 2014/2015 epidemic season implies the possibility of some mismatch between the genetic and antigenic features of the circulating virus and those included in the vaccine. That mismatch should not disenchant either from obtaining vaccination in the future seasons or from continuing efforts by savvy medical professional to increase vaccination coverage rate.

Acknowledgements This work was funded in parts by Specific Contract No 1-ECD.5235, implementing activities to the Framework Contract No/ECDC/2014/026, and NIPH-NIH thematic grant 5/EM.1

Conficts of Interest The authors declare no conflicts of interest in relation to this article.

References

Bednarska K, Hallmann-Szelińska E, Kondratiuk K, Brydak LB (2015) Evaluation of the activity of influenza and influenza like viruses in epidemic season, 2013–2014. Adv Exp Med Biol 857:1–7

Broberg E, Snacken R, Adlhoch C, Beauté J, Galinska M, Pereyaslov D, Brown C, Penttigen P, Buda S, Schweiger B (2015) Start of the 2014/2015 influenza season in Europe: drifted influenza A(H3N2) viruses circulate as dominant subtype. Euro Surveill 20 (4):1–5

Czarkowski MP, Hallmann-Szelińska E, Staszewka E, Bednarska K, Kondratiuk K, Brydak LB (2014) Influenza in Poland in 2011–2012 and in 2011/2012 and 2012/2013 epidemic season. Przegl Epidemiol 68 (3):455–463

Grohskopf LA, Olsen SJ, Sokolow LZ, Bresee JS, Cox NJ, Broder KR, Karron RA, Walter EB (2014) Prevention and control of influenza: recommendations of the Advisory Committee on Immunization Practices (ACIP) –United States, 2014–15 Influenza season. MMWR 63(32):691–697

Kissling E, Valenciano M, I-MOVE case–control studies team (2012) Early estimates of seasonal influenza vaccine effectiveness in Europe among target groups for vaccination: results from the I-MOVE multicentre case-control study, 2011/12. Euro Surveill 17(15): pii=20146

Valenciano M, Kissling E, Reuss A, Jiménez-Jorge S, Horváth JK, Donnell JMO, Pitigoi D, Machado A, Pozo F, I-MOVE Multicentre Case Control Study Team (2015) The European I-MOVE Multicentre 2013–2014 Case-control Study. Homogeneous moderate influenza vaccine effectiveness against A(H1N1) pdm09 and heterogeneous results by country against A (H3N2). Vaccine 33(24):2813–2822

Advs Exp. Medicine, Biology - Neuroscience and Respiration (2016) 24: 87
DOI 10.1007/5584_2016

Index

Printed by Printforce, the Netherlands